Monika Christina Maria Frey

Effects and mechanisms of a putative human pheromone

Monika Christina Maria Frey

Effects and mechanisms of a putative human pheromone

Androstadienone as putative modulator pheromone

Südwestdeutscher Verlag für Hochschulschriften

Impressum / Imprint

Bibliografische Information der Deutschen Nationalbibliothek: Die Deutsche Nationalbibliothek verzeichnet diese Publikation in der Deutschen Nationalbibliografie; detaillierte bibliografische Daten sind im Internet über http://dnb.d-nb.de abrufbar.

Alle in diesem Buch genannten Marken und Produktnamen unterliegen warenzeichen-, marken- oder patentrechtlichem Schutz bzw. sind Warenzeichen oder eingetragene Warenzeichen der jeweiligen Inhaber. Die Wiedergabe von Marken, Produktnamen, Gebrauchsnamen, Handelsnamen, Warenbezeichnungen u.s.w. in diesem Werk berechtigt auch ohne besondere Kennzeichnung nicht zu der Annahme, dass solche Namen im Sinne der Warenzeichen- und Markenschutzgesetzgebung als frei zu betrachten wären und daher von jedermann benutzt werden dürften.

Bibliographic information published by the Deutsche Nationalbibliothek: The Deutsche Nationalbibliothek lists this publication in the Deutsche Nationalbibliografie; detailed bibliographic data are available in the Internet at http://dnb.d-nb.de.

Any brand names and product names mentioned in this book are subject to trademark, brand or patent protection and are trademarks or registered trademarks of their respective holders. The use of brand names, product names, common names, trade names, product descriptions etc. even without a particular marking in this works is in no way to be construed to mean that such names may be regarded as unrestricted in respect of trademark and brand protection legislation and could thus be used by anyone.

Coverbild / Cover image: www.ingimage.com

Verlag / Publisher:
Südwestdeutscher Verlag für Hochschulschriften
ist ein Imprint der / is a trademark of
AV Akademikerverlag GmbH & Co. KG
Heinrich-Böcking-Str. 6-8, 66121 Saarbrücken, Deutschland / Germany
Email: info@svh-verlag.de

Herstellung: siehe letzte Seite /
Printed at: see last page
ISBN: 978-3-8381-3516-8

Zugl. / Approved by: Würzburg, JMU, Diss., 2012

Copyright © 2012 AV Akademikerverlag GmbH & Co. KG
Alle Rechte vorbehalten. / All rights reserved. Saarbrücken 2012

Herewith I thank my supervisors at the University of Würzburg and the Monell Chemical Senses Center in Philadelphia for their dedicated support of this project!

1 Content

1 Content	*1*
2 General introduction	*11*
2.1 What are pheromones?	13
2.2 Putative human pheromones	17
2.2.1 Natural occurrence of putative human pheromones	17
2.2.2 Androstenol, androstenone and estratetraenol	21
2.2.3 Androstadienone	23
2.2.4 Detection of putative human pheromones	27
3 Research questions	*29*
3.1 Does androstadienone affect attention-based reactions?	29
3.2 Does androstadienone affect behavioral tendencies?	30
3.3 Does androstadienone affect cortical reactions in women?	30
3.4 Does androstadienone affect cortical reactions in men?	31
4 Study I: attention and behavior	*32*
4.1 Introduction	32
4.1.1 Attention-based reactions	32
4.1.2 Attitudes and behavioral tendencies	37
4.2 Methods	38
4.2.1 Participants	38
4.2.2 Compounds and exposure	39
4.2.3 Material and apparatus	40
4.2.4 Procedure	41
4.2.5 Data analyses	42
4.3 Results	44
4.3.1 Odor ratings	44
4.3.2 Face ratings	44
4.3.3 Attention-based reactions	46
4.3.4 Attitudes and behavioral tendencies	47

4.4	Discussion	48
	4.4.1 Attention-based reactions	48
	4.4.2 Attitudes and behavioral tendencies	52
5	**Cortical reactions**	**56**
5.1	Introduction	56
5.2	Study II: cortical reactions in women	58
	5.2.1 Methods	59
	5.2.2 Results	70
	5.2.3 Discussion	90
5.3	Study III: cortical reactions in men	100
	5.3.1 Methods	100
	5.3.2 Results	105
	5.3.3 Discussion	111
6	**General discussion**	**116**
6.1	Androstadienone and explicit measures	116
6.2	Effect mechanisms of androstadienone	117
6.3	Contextual variables	119
6.4	Are androstadienone concentrations ecologically valid?	121
6.5	Androstadienone and the pheromone concept	123
	6.5.1 The pheromone concept in question	123
	6.5.2 Is androstadienone a human pheromone?	128
	6.5.3 Is androstadienone a human modulator pheromone?	132
6.6	Concluding remarks	133
7	**References**	**135**
8	**Appendix**	**160**
8.1	Material of study I	160
	8.1.1 Informed consent form	160
	8.1.2 Onscreen instructions	164
8.2	Material of study II	167

8.2.1	Informed consent form	167
8.2.2	Onscreen instructions	168
8.2.3	Plots of mood reactions	174
8.3	**Material of study III**	**179**
8.3.1	Informed consent form	179
8.3.2	Onscreen instructions	184
8.4	**Rating scales of study I and study III**	**187**
8.5	**Publications**	**188**
8.5.1	Articles	188
8.5.2	Conference proceedings	188
8.6	**Acknowledgment**	Fehler! Textmarke nicht definiert.

List of figures

Figure 1. The schematic angry and happy facial expressions used in the study 41

Figure 2. Mean reaction times in milliseconds for each odor and facial expression combination for both the push and the pull direction. Asterisk indicates the significant odor effect at $p < .05$ in response to angry faces. 47

Figure 3. Approach-avoidance score (±SEM) for each odor and participant's sex combination. Asterisks indicate significant differences of the approach-avoidance score from zero and between odors in men at $p < .05$. 48

Figure 4. Mean odor effects on feeling heavy ± SEM in control (dark) and androstadienone (light) groups. 75

Figure 5. Heart rate changes ± SEM of odor exposure in the control and the androstadienone groups. 76

Figure 6. Skin conductance changes ± SEM of odor exposure in the control and the androstadienone group. 77

Figure 7. Respiration rate changes ± SEM of odor exposure in the control and the androstadienone group. 77

Figure 8. Change scores in salivary cortisol levels in nmol/l after the first and second exposure to control and androstadienone 78

Figure 9. Mean P100 amplitudes ± SEM in response to happy, angry and neutral faces on electrode O2 in control (dark bar) and androstadienone groups (light bar). 79

Figure 10. Mean N170 amplitudes ± SEM in response to happy, angry and neutral faces on the electrode P8 in control (dark bar) and androstadienone groups (light bar). 80

Figure 11. Mean amplitudes ± SEM in response happy, angry and neutral faces 280 – 310 ms post stimulus on the electrode PO8 in control (dark bars) and androstadienone groups (light bars). 81

Figure 12. Mean amplitudes ± SEM on parietal sites (P3, Pz, P4) 320 – 360 ms post stimulus in response to happy, angry and neutral faces for the control and the androstadienone group. 82

Figure 13. Mean amplitudes ± SEM in response happy, angry and neutral faces 400 -600 ms post stimulus on Pz in control (dark bars) and androstadienone groups (light bars). 83

Figure 14. Mean P100 amplitudes ± SEM on O2 in response to positive, negative and neutral scenes of the categories couples, social and non-social in control (dark bars) and androstadienone groups (light bars). ... 84

Figure 16. Mean amplitudes ± SEM 280 – 310 ms post stimulus on PO8 in response to positive, negative and neutral scenes of the categories couples, social and non-social in control (dark bars) and androstadienone groups (light bars). 86

Figure 15. Mean amplitudes ± SEM 280 – 310 ms post stimulus in response to scenes in control (dark bars) and androstadienone groups (light bars). 86

Figure 17. Mean P300 amplitudes ± SEM on Pz in response to positive, negative and neutral scenes of the categories couples, social and non-social in control (dark bars) and androstadienone groups (light bars). ... 88

Figure 18. Mean amplitudes ± SEM 400 – 600 ms post stimulus on Pz in response to positive, negative and neutral scenes of the categories couples, social and non-social in control (dark bars) and androstadienone groups (light bars). 89

Figure 19. Mean amplitudes ± SEM 400 – 600 ms post stimulus on midline electrodes (Fz, Cz, Pz) in response to scenes with couples, social and non-social scenes, for the control (dark bars) and the androstadienone groups (light bars). 90

Figure 20. Presented cartoon faces with angry, happy and neutral expression 102

Figure 21. Mean P100 peaks ± SEM in response to happy, angry and neutral faces on O2 in control (dark bar) and androstadienone groups (light bar). 107

Figure 22. Mean N170 peaks ± SEM in response to happy, angry and neutral faces on P8 in control (dark bar) and androstadienone groups (light bar). 108

Figure 23. Mean EPN amplitudes ± SEM 220 – 260 ms post stimulus on P8 in response to happy, angry, and neutral faces in control (dark bar) and androstadienone groups (light bar). 109

Figure 24. Mean P300 amplitudes ± SEM 320 - 360 ms post stimulus on Pz in response to happy, angry and neutral faces in control (dark bar) and androstadienone groups (light bar). 110

Figure 25. Mean LPP amplitudes ± SEM 600 - 700 ms post stimulus on Pz in response to happy, angry and neutral faces in control (dark bar) and androstadienone groups (light bar). 111

Figure 26. Mean odor effects on feeling focused ± SEM in control (dark) and androstadienone (light) groups. 174

Figure 27. Mean odor effects on feeling social ± SEM in control (dark) and androstadienone (light) groups. 175

Figure 28. Mean odor effects on feeling energetic ± SEM in control (dark) and androstadienone (light) groups. 175

Figure 29. Mean odor effects on feeling open ± SEM in control (dark) and androstadienone (light) groups. 176

Figure 30. Mean odor effects on feeling relaxed ± SEM in control (dark) and androstadienone (light) groups. 176

Figure 31. Mean odor effects on the feeling of being sensual ± SEM in control (dark) and androstadienone (light) groups. 177

Figure 32. Mean odor effects on feeling irritated ± SEM in control (dark) and androstadienone (light) groups. 177

Figure 33. Mean odor effects on state anxiety ± SEM in control (dark) and androstadienone (light) groups. 178

List of tables

Table 1. Subjective intensity, pleasantness and familiarity ratings of androstadienone and the control solution (SD in brackets). 44

Table 2. Intensity and pleasantness ratings of angry and happy cartoon faces, while inhaling androstadienone or the control solution (SD in brackets) 46

Table 3. Experimental course of study II. 67

Table 4. Mean face ratings of androstadienone and control groups, SD in brackets 72

Table 5. Mean scene ratings of the androstadienone group, SD in brackets. 73

Table 6. Mean scene ratings of the control group, SD in brackets 74

Table 7. Mean reaction times of participant's median reaction times of tonic and phasic alertness values in androstadienone and control groups in ms, SD in brackets. 75

Table 8. Intensity, pleasantness and familiarity ratings of androstadienone and the control solutions (SD in brackets)................ 105

Table 9. Pleasantness and intensity ratings of angry, happy and neutral cartoon faces, while smelling androstadienone or control solution (SD in brackets) 106

2 General introduction

The power of smell undeniably influences our everyday lives more than we are aware of. Odors of all different kinds evoke memories and emotions, influence our dreams, affect our mood and enable us to enjoy eating and drinking or to detect dangerous situations, even on a subconscious level (reviewed in Doty, 2003a). The value of our sense of smell gets more obvious if we lose it. Anosmic people describe the world as poor and grey, lacking an unconscious background for everything else. Moreover, the sense of smell is important for mate choice and the survival of our species. Recent research has shown that human sperms express a functioning olfactory receptor, which may be crucial for the fertilization process (Spehr et al., 2003). However, for long time the importance of our sense of smell was misjudged and thought to play a minor, unimportant role in our behavior (Herrick, 1924). Based on the relative size of the olfactory brain structures Broca divided mammals into macrosmatics and microsmatics (Broca, 1888). Microsmatics, like primates or whales, have a weak sense of smell or poorly developed olfactory organs. Humans are also labeled as microsmatics because of the relatively small size of the olfactory cortex compared to that of other species, although the absolute sizes do not necessarily differ between species. However, olfactory detection threshold values for aliphatic esters are comparable between rats, labeled as macrosmatics and humans (Salazar, Laska, & Luna, 2003). Studies comparing the olfactory discrimination ability in humans and monkeys demonstrated that neither genetic nor anatomical features are reliable predictors for olfactory performance. Although monkeys possess twice as many functional olfactory receptor genes and a relatively larger olfactory cortex than humans, discrimination performance showed similar patterns in both species (Laska & Freyer, 1997; Laska, Genzel, & Wieser, 2005). Furthermore, humans are also able to discriminate among thousands of different airborne substances, even in extremely low concentrations. Our sense of smell is therefore highly complex and sensitive. Besides, olfaction and cognition are closely linked; olfactory impulses are directly causing cognitive processes via projections to the thalamus, amygdala, entorhinal cortex and hypothalamus. This indicates to a significant influence of our sense of smell on various body reactions.

One of the most important olfactory cues is our body odor. Among primates, humans have the most and largest sweat producing apocrine glands, which make us to the "most highly scented ape of all" (p. 270) (Wyatt, 2003). These scents, the familiar body odors of relatives and progeny or the surprisingly infatuating scent of a stranger passing by, are highly complex and probably individually distinctive (Schaal & Porter, 1991). That we place high prominence on our own scent attests the multi-billion dollar perfume industry. We all try to improve our natural body odors. Before another person's visual appearance can fascinate and attract us, the female nose especially must become completely infatuated with the novel encounter. But why does anybody perceive a different person's smell as most likable and how is this preference mediated? To find answers, the scientific world tries to confirm influences of odors on our social life. And indeed, the preference for specific body odors has a more important function than simply identifying them as either nice and pleasant or unpleasant. An inherent influence on odor preferences has been determined in that our odor bias is related to the major histocompatibility complex (MHC) composition. The MHC is a highly variable and conserved set of genes related to our immune system. Milinski and colleagues (2001) found that genetically similar subjects prefer similar scents for themselves, suggesting that persons choose a perfume that amplifies their own body odor and coevally communicates information about their immunogenetics (Milinski & Wedekind, 2001). In contrast, in the light of human mate choice the nose prefers genetically dissimilar partners to account for the best possible genetic combination in order to achieve the largest benefit for the offspring (Havlicek & Roberts, 2009).

Is it possible that our social life is influenced by undetectable chemicals, called pheromones? Possibly all non-human animals, vertebrates and invertebrates, use chemical signals to communicate about food, territory and sex. The idea that humans might be similarly influenced has been controversially discussed among scientists. Putative human pheromones, like androstadienone, are supposed to have special modulator functions on psychological as well as physiological and behavioral reactions. However, humans as the most complex organisms on earth, with highly developed cognitive functions, are influenced by many factors like environmental and physical conditions, emotions and social interactions, all of which have to be integrated and evaluated. To provide a further step towards elucidating interactions between different modalities especially in a social context, the present thesis explores the influence of a social olfactory cue, the sweat compound androstadienone, on social visual stimuli and behavioral reactions resulting from this interaction. Specifically, mechanisms underlying psycho physiological and behavioral effects, which result from inhaling the endogenous odorant, are addressed.

2.1 What are pheromones?

Almost eighty years ago, the entomologist Albrecht Bethe described "endohormones" as hormones secreted within the body and "ectohormones" as hormones secreted outside of the body in insects, dividing the latter into agents working intraspecifically (homoiohormones) and interspecifically (alloiohormones) (Bethe, 1932). Karlson and Lüscher (1959) then redefined the term "homoiohormones" into pheromones from the Greek *pherein* "to transfer" and *hormon* "to exite". This new term was supposed to highlight the difference between hormones, which are produced by endocrine glands. This new group of chemical compounds was defined as "substances which are secreted to the outside by an individual and received by a second individual of the same species, in which they release a specific reaction, for example a definite behavior or a developmental process" (Karlson & Lüscher, 1959). The first pheromone bombykol was isolated from the silkworm moth *Bombyx mori*, which is its sexual attractant. The female produces an extremely low concentration of bombykol, only about 200 molecules. The male moth is able to follow the discontinuous gradient and finally finds his potential mating partner. Since then, pheromones have been found all across the animal kingdom, in the first instance in insects but also in land mammals as well as fish and underwater crustaceans. Also yeast, ciliates, algae and bacteria use these substances to send messages to conspecifics. There are alarm pheromones in aphids, aggregation pheromones in ants and territorial pheromones in dogs, to mention only a few (Wyatt, 2009).

Pheromones are classically divided into two classes based on their functionality in insects: releaser and primer pheromones. Releaser pheromones act from seconds to minutes and stimulate a specific behavioral reaction, e.g. aggregation, trail following or sexual attraction. One example of a mammalian releaser pheromone is 5α-androst-16-en-3-one (androstenone). It is produced by the boar's testes and elicits a freezing behavior in the sow to enable successful mating. This phenomenon is commercially used for artificial insemination (Melrose, Reed, & Patterson, 1971). In contrast, primer pheromones do not elicit a specific behavioral reaction, but induce a physiological change in the receiver. Acting over a longer time span like hours or days they influence the hypothalamic-pituitary-gonadal axis, which in turn alters hormone levels and changes behavioral responses (Kohl, Atzmueller, Fink, & Grammer, 2001; Wilson & Bossert, 1963). Well known effects of primer pheromones in mammals are the Vandenbergh-Effect, where the first ovulation in female mice starts earlier with exposure to a substance of male's urine (Vandenbergh, 1969) and the Bruce-Effect, where a fertilized egg does not nest if the female mouse is exposed to an unknown male (Bruce, 1959). Primer and releaser effects of pheromones are not mutually exclusive, because endocrine responses like cortisol changes might in turn affect stressful behavior and vice versa. But specific behaviors in mammals are released rarely. Thus, to fit the term releaser pheromone more into mammalian reality pheromonologists morphed it into the terms signaling pheromone, informer pheromone, modulator pheromone or behavioral pheromone. Signaling pheromones were defined as to communicate complex information about the sender. This might reach from body condition to maturation, social or hierarchical status (Johnston, 1998). Rodents, for example, have the ability to discriminate relatives from non-relatives by their urine's odor type. This is mediated via the MHC, which are genes strongly linked to immune function. To choose mates with a different MHC and therefore with a different body odor prevents inbreeding, homozygosity and abortion (Beauchamp & Yamazaki, 2003; Yamazaki, Beauchamp, Curran, Bard, & Boyse, 2000; Yamazaki, Curran, & Beauchamp, 1999).

Although, there are hundreds of studies claiming pheromone effects in mammals, only very few putative pheromonal substances have been isolated. Moreover, the term pheromone means different things to different people and has been redefined in various attempts to fit a range of chemical substances eliciting behavioral and endocrine functions into a common characterization. For example, the highly complex odors used in mammals to distinguish strangers and relatives, do not fit into the classical definition of insect pheromones, being a single compound rather than mixtures. Beauchamp et al. (1976) therefore proposed operational requirements for mammalian pheromones to distinguish between olfactory and pheromonal responses: the compounds have to be species-specific, elicit a well-defined behavior or endocrine function, are dependent on a large degree of genetic programming, consist of one or only a few compounds and produce a unique behavioral or endocrine response not demonstrated by other similar stimuli. Buck (2000) redefined mammalian pheromones as eliciting "programmed neuroendocrine changes and innate behaviors" suggesting the need for a very precise recognition process (Buck, 2000). To apply the pheromone concept to humans is even more complicated. Preti and Wysocki (1999) pointed out that human behavior is influenced by many factors. Rather than eliciting a definite and immediate response, human pheromones might modulate the likelihood and intensity of an individual response (Preti & Wysocki, 1999). Another definition stressed the communicational aspect of pheromones rather than that of pheromonal chemicals. Meredith and colleagues (2001) defined pheromones as "chemical substances released by one member of a species as communication with another member, for their mutual benefit" (Meredith, 2001). The criteria of mutual benefit should help to clarify the development of communication via chemical substances. In addition, the observation that human sweat is able to modulate endocrine functions in women led to following definition: "Pheromones are airborne chemical signals that are released by an individual into the environment and which affect the physiology or behavior of other members of the same species" (p. 177, Stern & McClintock, 1998). Then, based on psychological effects of the isolated endogenous odorant androstadienone, a new pheromone class was suggested: the modulator pheromones (Jacob & McClintock, 2000). An official definition was proposed by McClintock in 2003: "Modulator pheromones modulate ongoing behavior or a psychological reaction to a particular context, without triggering specific behavior or thoughts. They change stimulus sensitivity, salience and sensory-motor integration" (McClintock, 2003).

2.2 Putative human pheromones

2.2.1 Natural occurrence of putative human pheromones

Human body odors communicate social information, like emotional states, gender, mate value or degree of kinship (Chen & Haviland-Jones, 2000; Penn et al., 2007; Rikowski & Grammer, 1999; Roberts et al., 2005; Wallace, 1977; Wedekind, Seebeck, Bettens, & Paepke, 1995; Weisfeld, Czilli, Phillips, Gall, & Lichtman, 2003). More importantly, several studies provide evidence that human axillary secretions modulate endocrine functions, perception, cognition and behavior (Chen, Katdare, & Lucas, 2006; Pause, Adolph, Prehn-Kristensen, & Ferstl, 2009; Prehn, Ohrt, Sojka, Ferstl, & Pause, 2004). Therefore, it has been suggested that human body odors contain specifically active compounds, which can be considered human pheromones.

Human body odors themselves are processed specifically by our brain. A functional imaging study showed that a stranger´s body odor in contrast to a similar common odor activates emotionally relevant brain areas like the amygdala and insular region (Lundström, Boyle, Zatorre, & Jones-Gotman, 2008). Furthermore, the study proved that body odors, as ecologically relevant stimuli, are processed in specialized networks distinctly separate from the common olfactory system. Especially sexual sweat, conveying socio-emotional information, activates brain areas which are not only involved in olfactory processing but also in emotional and social processing like the hypothalamus and the fusiform gyrus (Zhou & Chen, 2008b). Moreover, humans are able to distinguish between self and non-self odors at a very early stage of stimulus processing. Event-related potentials showed that the personal odor is processed faster than the body odor of a stranger (Pause, Krauel, Sojka, & Ferstl, 1998).

Even more interesting in terms of our social life is the connection between the exposure to human sweat and the perception of another person. Rikowski and Grammer (1999) found a significant positive correlation between the judgment of men´s facial attractiveness and sexual attractiveness ratings of their body odors by women. With exposure to male´s axillary secretions, women rated male faces as more attractive (Thorne, Neave, Scholey, Moss, & Fink, 2002). Moreover, body odors have different effects depending on their source and composition. Participants smelling sweat, collected from feared donors, are more likely to perceive an ambiguous face as fearful than participants smelling a control odor (Zhou & Chen, 2008a). In addition, fear sweat sharpens emotional face recognition and elicits a specific activation in the amygdala (Mujica-Parodi *et al.*, 2009). These results show that different modalities like olfaction and vision are integrated in the human brain and that especially social chemosensory signals have the ability to affect central nervous reactions.

The first hint that human sweat contains substances that are able to modulate endocrine functions, defined in animals as primer pheromones, was published by Martha McClintock (McClintock, 1971). She found that women living together in a dormitory develop a synchronized menstrual cycle, which is a similar phenomenon known as the Whitten-effect found among rodents (Whitten, Bronson, & Greenstein, 1968). A preliminary study supported this finding and demonstrated that female underarm sweat swiped on the upper lip of women can also shift the menstrual cycle of the receiving women towards the donor´s cycle (Russell, Switz, & Thompson, 1980). Extending these results, a well-controlled laboratory study found that women had shorter cycles if they were exposed to female sweat collected during the follicular phase. The opposite, i.e. a longer menstrual cycle, occurred if women were exposed to substances collected during the ovulary phase (Stern & McClintock, 1998). An androgen substance has been implicated to cause these effects. 5α-andorst-16-en-3α-ol (androstenol, which will be further reviewed below) applied to women´s upper lip indeed decreased the luteinizing hormone (LH) surge in women, suggesting a regulatory influence on the ovulary timing in humans (Shinohara, Morofushi, Funabashi, Mitsushima, & Kimura, 2000). The hypothesis that women emit volatile substances affecting endocrine functions was further supported by female axillary sweat changing the frequency of LH pulses in women (Shinohara, Morofushi, Funabashi, & Kimura, 2001).

More single compounds found in human sweat have been suggested to act as human pheromones. Their production is primarily linked to the apocrine glands of the human skin. In contrast to the thermo regulating eccrine sweat glands, which are autonomously activated during exercise and stress, the apocrine glands produce odorless sweat while sexually aroused or while being in other emotional states (Wilke, Martin, Terstegen, & Biel, 2007). Especially fear and anxiety are closely functionally linked to these glands; therefore, authors supposed that they represent a vestigial defense system. The highest density of apocrine glands is in the axillae and the perineum (Doty, 1981). Apocrine secretions contain 16-androstenes, which are metabolized in human testes through testosterone. The most prominent 16-androstenes found in fresh human sweat of axillae treated with diethyl ether to prevent bacterial activity are 4,16-androstadien-3-one (androstadienone) with 17.9 pmol/cm² and androstenol with 6.9 pmol/cm² (Gower, Holland, Mallet, Rennie, & Watkins, 1994). It is important to note that these extracts in the mentioned concentrations are almost odorless. The characteristic sweat odor mainly arises through microorganisms, the aerobic bacteria Corynebacterium ssp., which transform the precursors androstadienol and androstadienone into the urine like odorant 5α-androst-16-en-3-one (androstenone) (Claus & Alsing, 1976; Gower et al., 1994). Men produce about five times higher concentrations of androstenone than women (Brooksbank, Wilson, & MacSweeney, 1972; Gower, Bird, Sharma, & House, 1985). This is due to different blood levels of androgens as well as different characteristics of bacterial skin colonization. Male´s axillae are dominated by the aforementioned bacteria Corynebaceria ssp., whereas female´s axillary skin microflora contains mainly the bacteria Micrococcacea (Jackman & Noble, 1983).

Interestingly, these bacteria are only present in human axillae after the apocrine sweat glands reach maturity during puberty (Stoddart, 1990). Most studies, aiming to find pheromonal effects of human sweat substances, focused on the three mentioned 16-androstenes: androstenone androstenol and androstadienone. Furthermore, the putative female pheromone 1,3,5(10),16-estratetrael-3-ol (estratetraenol) has also been under favored investigation. This estrogen was identified with approximately 100 μg/liter of urine of pregnant women (Thysen, Elliott, & Katzman, 1968).

2.2.2 Androstenol, androstenone and estratetraenol

One of the first reports of pheromonal effects of an isolated substance in humans was published by Cowley (1977). Women wearing a surgical mask impregnated with androstenol judged men as more favorable, but not women (Cowley, Johnson, & Brooksbank, 1977). Subsequent research following this approach found similar effects such that androstenol treated male and female participants rated pictures of women as more sexually attractive and pictures of men and women as warmer (Kirk-Smith, Booth, Carroll, & Davies, 1978). Androstenol also increased male´s attractiveness ratings of a target male but not of a target female (Filsinger, Braun, & Monte, 1985). Moreover, women exposed to androstenol rated their moods as submissive rather than aggressive during ovulation, which might be due to sensitivity changes during the menstrual cycle (Benton, 1982). However, inhaling androstenol once a day did not influence women´s sexual arousal or mood (Benton & Wastell, 1986).

In contrast to a no-odor condition, the exposure to androstenol and androstenone decreased female´s sexual attractiveness ratings of the target man and decreased self-rated sexual attractiveness in male participants (Filsinger et al., 1985). Faces were rather perceived as more masculine as a result of inhalation of androstenone compared to estratetraenol or water (Kovacs *et al.*, 2004). Interestingly androstenone influenced participant´s chair preference in a dental office waiting room. Women were more likely to sit in a chair odorized by androstenone than in chairs not sprayed with the odor (Kirk-Smith & Booth, 1980). Interestingly, around the time of highest fertility, compared to other menstrual cycle phases, women perceive the smell of androstenone as less aversive (Grammer, 1993). However, sensitivity to androstenone and androstenol is highly variable and due to sensitization, experience, age, sex, sexual orientation, genetic determination and menstrual cycle phases (Boyle et al., 2006; Bremner, Mainland, Khan, & Sobel, 2003; Dorries, Schmidt, Beauchamp, & Wysocki, 1989; Keller, Zhuang, Chi, Vosshall, & Matsunami, 2007; Knaapila et al., 2008; Lubke, Schablitzky, & Pause, 2009; Morofushi, Shinohara, Funabashi, & Kimura, 2000; Wysocki & Beauchamp, 1984; Wysocki, Beauchamp, Schmidt, & Dorries, 1987).

Studies exposing human subjects to estratetraenol showed sex-specific and context-dependent effects on mood, physiology and brain activation. Women reported an increase in positive mood, whereas men reported a decrease in positive mood state (Jacob & McClintock, 2000). Furthermore, an increased physiological arousal, i.e. lower skin temperature and higher skin conductance, was found in women but not in men following estratetraenol exposure (Jacob, Hayreh, & McClintock, 2001). However, another study has failed to reproduce these effects (Bensafi *et al.*, 2003). But research proved a contextual dependence of estratetraenol effects. In both men and women, estratetraenol increased sexual arousal only in a sexual context (Bensafi, Brown, Khan,

Levenson, & Sobel, 2004). Furthermore, activity changes in the hypothalamus were described in men, but not in women (Savic, Berglund, Gulyas, & Roland, 2001).

2.2.3 Androstadienone

This thesis focuses on the endogenous compound androstadienone. This substance has been singled out as the most likely candidate of a human pheromone. It is found in human sweat (Gower et al., 1994), axillary hair (Nixon, Mallet, & Gower, 1988), male testes and blood plasma (Brooksbank et al., 1972). As will be reviewed below, several studies reported that androstadienone affects various human responses compared to even structurally and perceptually similar odors.

2.2.3.1 Androstadienone effects on physiological and psychological reactions

One of the first reports about androstadienone effects was published by Grosser and colleagues (2000). Women responded to androstadienone, directly applied to the vomeronasal organ, with reduced nervousness, tension and other negative feeling states as well as changes in autonomic physiology (Grosser, Monti-Bloch, Jennings-White, & Berliner, 2000). These findings were replicated several times. An increase in physiological arousal, as indicated by increased heart rate, skin conductance, blood pressure, respiration rate and a decreased skin temperature, was found in women, but only if a man was present (Jacob, Hayreh et al., 2001; Lundström & Olsson, 2005). In contrast androstadienone increased men's skin temperature, i.e. a reduced physiological arousal in men. This was unaffected by the socio-experimental condition, i.e. if a man was present or not (Jacob, Hayreh et al., 2001). However, in a sexually arousing context androstadienone increased the skin temperature in both sexes left alone during testing (Bensafi, Brown et al., 2004).

Whereas androstadienone maintains (Bensafi, Brown et al., 2004; Jacob, Garcia, Hayreh, & McClintock, 2002; Jacob & McClintock, 2000) or even enhances (Lundström & Olsson, 2005; Villemure & Bushnell, 2007) positive mood in women, men reported a decrease in positive emotion (Jacob & McClintock, 2000). In a negative context it increases the perceived heat pain intensity especially in women, suggesting an increased attention towards emotionally negative stimuli (Villemure & Bushnell, 2007). Furthermore androstadienone maintains higher levels of cortisol in women (Wyart *et al.*, 2007).

2.2.3.2 Androstadienone effects on the human brain

That the human brain reacts differently to androstadienone compared to common odors was demonstrated by different brain imaging techniques. By using chemosensory event-related brain potentials it has been documented that androstadienone is processed more rapidly and automatically than the structurally similar androstenone or structurally dissimilar H_2S in the female brain (Lundström, Olsson, Schaal, & Hummel, 2006). This is a hint for its pheromonal properties in such that androstadienone might be evolutionarily more significant and therefore processed preferentially in an olfactory subsystem. Grosser and colleagues (2000) found increased alpha activity during androstadienone exposure. Alpha waves are electromagnetic oscillations produced during relaxation with open or closed eyes. These waves seem to be influenced by androstadienone through tension-reducing hypothalamic activation. This was confirmed later by brain imaging studies, which demonstrated a sex-specific activation of the hypothalamus by smelling androstadienone in heterosexual women but not in heterosexual men (Frasnelli, Lundström, Boyle, Katsarkas, & Jones-Gotman, 2008; Savic et al., 2001). These observations were further extended by comparing brain responses to androstadienone in participants with different sexual orientations. In contrast to heterosexual men, homosexual men displayed the same hypothalamic activation as it was found in heterosexual women (Savic, Berglund, & Lindström, 2005). The hypothalamus was also significantly activated with androstadienone compared to estratetraenol in non-homosexual male-to-female transsexuals (Berglund, Lindström, Dhejne-Helmy, & Savic, 2008). As these hypothalamic areas are involved in sexual behavior (Oomura, Aou, Koyama, Fujita, & Yoshimatsu, 1988) androstadienone seems to play a critical role in human sexual behavior.

Androstadienone also affects brain areas associated with emotional and attentional processing. A study using positron emissions tomography (PET) found an activated superior temporal cortex (STP) and the fusiform gyrus with exposure to androstadienone but not to pleasant, unpleasant and neutral control odors in women (Gulyas, Keri, O'Sullivan, Decety, & Roland, 2004). Both the STP and the fusiform gyrus are involved in the recognition of facial features and emotional expressions (Haxby, Hoffman, & Gobbini, 2000), thus indicating specific effects of androstadienone on face processing. Androstadienone also alters cerebral glucose utilization in areas like the prefrontal cortex, the cingulate cortex and the amygdala, which are assumed to process emotional stimuli (Jacob, Kinnunen, Metz, Cooper, & McClintock, 2001). Based on the latter findings the authors suggested that androstadienone might specifically influence the processing of visual stimuli with emotional content via projections from the amygdala (Amaral, Price, Pitkanen, & Carmichael, 1992). Concordantly, activity changes in areas like the occipital and the parietal cortex suggested an influence of androstadienone on attentional processes in women (Jacob, Kinnunen et al., 2001).

2.2.3.3 Androstadienone effects on behavior

Androstadienone also affects human behavior in association to mate choice. The first support for this link originated from Cornwell and colleagues (2004), who found a positive correlation between women´s preference for masculine faces as a long term mate and the pleasantness of androstadienone (Cornwell *et al.*, 2004). A study, examining androstadienone functioning in a sexually significant context, was conducted in an ecologically valid environment. After a speed-dating event, women, who had been exposed to androstadienone, rated their male interaction partners as more attractive than women, who had been exposed to clove oil or water (Saxton, Lyndon, Little, & Roberts, 2008). Furthermore, three consecutive studies showed androstadienone effects on implicit behavioral measurements (Hummer & McClintock, 2009). Participants had to conduct dot-probe and stroop tasks with faces and words, respectively. Androstadienone enhanced participant´s attention specifically to emotional information. These effects were independent of participant´s gender and of social or non-social value of this information.

Taken together, biopsychological responses to androstadienone are obviously dependent on environmental conditions and individual characteristics. Androstadienone affects reactions in both, men and women, sometimes in a sex-specific way. Reported findings are hints for androstadienone to be an evolutionarily significant stimulus with pheromonal properties. Preferentially processed in an olfactory subsystem, it modulates behavior by psychological mechanisms on a subconscious level. The broad range of results asks for a clarification of the underlying mechanisms through which androstadienone is working.

2.2.4 Detection of putative human pheromones

In most reptiles, amphibians and mammals pheromones are processed in an accessory olfactory system, with the vomeronasal organ (VNO) as the peripheral signal organ. This organ is housed in the nasal cavity and is of highest importance for pheromonal communication with conspecifics (Kimchi, Xu, & Dulac, 2007). In human adults, the existence and functionality of a VNO is highly questioned. Anatomical studies have found the VNO, a small opening with a diameter of 1.0 - 2.5 mm, uni- or bilaterally, in 25% - 100% of human subjects depending on the endoscopic method and reported a change in visibility over time (Gaafar, Tantawy, Melis, Hennawy, & Shehata, 1998; Johnson, Josephson, & Hawke, 1985; Knecht, Kuhnau, Huttenbrink, Witt, & Hummel, 2001; Stensaas, Lavker, Montibloch, Grosser, & Berliner, 1991; Trotier et al., 2000). Positive evidence for the functionality was published almost exclusively by Monti-Bloch and colleagues (Berliner, Monti-Bloch, Jennings-White, & Diaz-Sanchez, 1996; Monti-Bloch, Diaz-Sanchez, Jennings-White, & Berliner, 1998; Monti-Bloch & Grosser, 1991; Monti-Bloch, Jennings-White, & Berliner, 1998; Monti-Bloch, Jennings-White, Dolberg, & Berliner, 1994; Stensaas et al., 1991; Takami et al., 1993). They found that estratetraenol and androstadienone activates the VNO in a sex-specific way. The female VNO was only activated if directly stimulated with androstadienone, whereas the male VNO was only activated if stimulated with estratetraenol. Measured VNO cell potentials did not rise with an olfactory control stimulus or if measured on the nasal respiratory epithelium (Monti-Bloch & Grosser, 1991). However, these studies have been highly criticized, not least because none of them have successfully been replicated independently. More recent research, however, suggests that putative human pheromones are not necessarily processed by the VNO. Estratetraenol, for example, elicits a hypothalamic activation in healthy men, but not in male patients suffering from nasal polyps (Savic, Heden-Blomquist, & Berglund, 2009). This nasal congestion prevented odors to reach the main olfactory epithelium, but not the VNO. This has been taken as evidence that putative pheromones are perceived via the main olfactory system. In line, the occlusion of the VNO with a latex patch did not affect the threshold of androstenone (Knecht *et al.*, 2003). More importantly for the current thesis androstadienone does produce typical brain activation, although the VNO had been functionally occluded or was even absent (Frasnelli et al., 2008). Also the perception of or sensitivity to androstadienone was not influenced by functional occlusion or absence of the VNO. Authors concluded that the VNO in humans had no obvious function, at least no significance in processing androstadienone. The existence of a human odorant receptor, which is expressed in the human nasal epithelium, but not in the VNO and selectively responds to androstenone and androstadienone, further supports this assumption (Keller *et al.*, 2007).

3 Research questions

The primary function of our senses is to provide information about our environment. All inputs are integrated into a complex cognitive picture. The sense of smell and the sense of sight are fused in the brain influencing each other and leading to a higher-order construct, which, in turn, results in specific behavioral responses. The over-reaching aim of this thesis was to explore whether androstadienone modulates individual´s reactions to visual stimuli. This issue was subdivided into four research questions, which were addressed by three separate empirical studies. The first experiment investigated androstadienone effects on behavioral reactions. Study II and III explored the effects of androstadienone on brain reactions to visual stimuli.

3.1 Does androstadienone affect attention-based reactions?

One recurrent explanation for androstadienone effects is that it may modulate attentional processes. Behavioral and neurophysiological studies reported an influence on general attention (Jacob, Kinnunen et al., 2001; Lundström, Goncalves, Esteves, & Olsson, 2003), as well as an improvement of attention specifically towards affective information (Hummer & McClintock, 2009; Villemure & Bushnell, 2007). The first experiment aimed at exploring possible effects of androstadienone on attention-based motor reactions towards social emotional stimuli. The approach-avoidance task was used to reveal information about how accurately and quickly humans evaluate and consecutively react to happy and angry faces with specific arm movements, while exposed to minute amounts of androstadienone or a control solution.

3.2 Does androstadienone affect behavioral tendencies?

Androstadienone is known to improve the evaluation of our conspecifics. Therefore, it may modulate our willingness to establish or maintain social relationships, which might result in specific behavior. However, so far no studies have explored behavioral tendencies resulting from previous appraisal of incoming stimuli with respect to androstadienone exposure. Whether androstadienone modulates approach or avoidance tendencies towards emotional faces was tested with the approach-avoidance task.

3.3 Does androstadienone affect cortical reactions in women?

Women´s brain activation by androstadienone is known to differ from activation by common odors (Savic et al., 2001; Savic et al., 2005). Especially, an activation alteration by androstadienone of regions associated with visual processing and social cognition was reported (Gulyas et al., 2004). Therefore, study II explored whether androstadienone influences the central nervous processing of visual stimuli. A common assumption is that androstadienone may influence attentional processes. Event related potentials (ERPs) recorded by electroencephalography, reflecting allocation of attentional resources in the brain, should clarify whether androstadienone acts via attentional mechanisms. Moreover, it has been suggested that an appropriate social or emotional context is necessary to detect androstadienone effects. To clarify this issue the processing of different picture categories and valences was tested, while either exposed to androstadienone or a control solution.

3.4 Does androstadienone affect cortical reactions in men?

Results from study I indicate androstadienone effects on attention-related reactions towards angry faces in men and women. Attention allocation to emotional faces has also been suggested by Hummer and McClintock (2009). Study II tentatively indicates an androstadienone related effect on central nervous face processing in women. Therefore, study III investigated whether androstadienone affects the central nervous face processing of men.

4 Study I: attention and behavior

4.1 Introduction

4.1.1 Attention-based reactions

Faces, among the most important visual stimuli for social interactions, are detected and encoded at a very early stage of processing in the brain (Eimer, 2000; Eimer & McCarthy, 1999; Linkenkaer-Hansen et al., 1998; Mühlberger et al., 2009). Arguably, discriminating between friend and foe, i.e. between friendly or threatening interaction partners, is a vital function of our early visual detection system. Indeed, it has been demonstrated that the brain is specifically attuned to detect angry faces at an early stage of the visual processing stream and that these so-called fear-related visual stimuli are tagged for preferred processing (Schupp et al., 2004). Preferential attention to angry faces has also been documented in visual search paradigms. Faster and more accurate detection of angry compared to happy target faces was interpreted as a threat advantage influenced by the amygdala and the associated fear module (Esteves, Parra, Dimberg, & Öhman, 1994; Fox, Russo, Bowles, & Dutton, 2001; Fox, Russo, & Dutton, 2002; Öhman, Lundquist, & Esteves, 2001; Öhman & Mineka, 2001). These effects are often interpreted as angry faces signaling a possible threat to the observer. Overall, the majority of research indicates a preattentive processing and triggering of fear reactions to threatening stimuli, such as angry faces (Eastwood, Smilek, & Merikle, 2001; Hansen & Hansen, 1988; Öhman, 2007; Williams & Mattingley, 2006).

Furthermore androstadienone and human body odors, which is the natural source of androstadienone, activates fear related brain areas, i.e.

the amygdala and insular region, in contrast to a similar common odor (Jacob, Kinnunen et al., 2001; Lundström et al., 2008). These results were extended by studies exposing human subjects to sweat collected from emotionally stressed donors. Participants were more likely to perceive an ambiguous face as fearful when they smelled fear sweat compared to a control odor (Zhou & Chen, 2008a). It has been suggested that fear sweat contains alarm pheromones which are processed faster in the brain and act via the amygdala. Olfactory fear signals might therefore interfere with the perception of ambiguous visual fear stimuli and fasten their recognition. A recent brain imaging study replicated these results and verified the hypothesis: fear sweat indeed sharpened emotional face recognition and furthermore elicited a specific activation in the amygdala (Mujica-Parodi et al., 2009).

Recent studies have indicated that endogenous olfactory information can also trigger fear-like cortical processing. Exposure to sweat from an unknown person triggers the amygdala and insular cortex (Lundström et al., 2008), two main components of the cerebral fear network (Morris, Öhman, & Dolan, 1998; Whalen et al., 1998). Also, body odors sampled from individuals in an emotionally arousing setting, e.g. the first sky dive, have been shown to activate the amygdala and the insula, thus indicating that chemosensory signals sampled from humans can trigger the brain's fear processing system (Mujica-Parodi *et al.*, 2009; Prehn-Kristensen *et al.*, 2009). Interestingly, Lundström and colleagues (2008) demonstrated that smelling body odors elicited a strong activity in the occipital cortex independent of behavioral task or visual input, which suggests that exposure to body odors might elicit preparedness in the visual system for a potential encounter.

In line, recent research demonstrated that exposure to the endogenous odorant androstadienone a compound of human body odors (Gower et al., 1994) indeed modulates participants' responding to visual stimuli (Hummer & McClintock, 2009). Furthermore, this human compound has recently been suggested to be a potential human pheromone (Sobel & Brown, 2001). Supporting this assumption, several studies have demonstrated androstadienone effects on psychological and physiological variables in men and women. Inhalation of minute amounts of androstadienone prevented a drop in positive mood as well as an increase in negative mood in male and female participants (Jacob *et al.*, 2002). In addition, men and women reacted with increased ear pulse rate, higher skin temperature and increased self rated sexual arousal during androstadienone presentation compared to a common control odor (Bensafi, Brown et al., 2004). Other studies found sex-specific androstadienone effects such that women showed an increase in positive mood and physiological arousal (Jacob, Hayreh et al., 2001; Jacob & McClintock, 2000; Lundström & Olsson, 2005; Villemure & Bushnell, 2007), whereas men responded to androstadienone with a decrease in positive emotions and in physiological arousal (Jacob, Hayreh et al., 2001; Jacob & McClintock, 2000). Furthermore androstadienone effects have been shown to depend on the emotional context (Bensafi, Brown et al., 2004). After inducing sadness, but not sexual arousal or happiness, androstadienone increased negative mood in men and kept positive mood in women. Moreover, androstadienone maintained higher levels of cortisol and increased perceived intensity of a pain stimulus in women (Villemure & Bushnell, 2007; Wyart et al., 2007). The latter result indicates that androstadienone is able to affect attention allocation in this case specifically towards a negative stimulus.

This hypothesis is supported by studies using brain imaging and psychological measures. Activity changes in brain areas like the occipital and the parietal cortex indeed suggest an influence of androstadienone on attentional processes (Jacob, Kinnunen et al., 2001). As mentioned above, participant´s performance in implicit visual attention tasks was modulated such that attentional resources were more engaged in the processing of emotional compared to neutral stimuli while man and women were exposed to androstadienone (Hummer & McClintock, 2009). Furthermore, the endogenous odor enhanced the subjective feeling of paying attention in men and women (Jacob, Hayreh et al., 2001; Lundström et al., 2003). Remarkably, one field experiment showed behavioral androstadienone related effects (Saxton et al., 2008): women exposed to androstadienone compared to women exposed to a control odor rated the man whom they met during a speed dating event as more attractive. Taken together, empirical evidence supports the view that androstadienone has the capacity to influence humans' physiology, mood and behavioral responses.

As outlined above, it has been well established that visual stimuli with high behavioral relevance, like angry faces, receive preferential processing. Moreover, body odors that contain androstadienone activate the human fear system in the brain. Since these endogenous odors are omnipresent in close social interactions, we hypothesized that androstadienone would fasten the processing of angry faces and in turn facilitate behavioral responses to these stimuli. Additionally, motivational tendencies, like approach or avoidance reactions, in response to emotional faces might be affected by androstadienone. Because body odors, especially from an ominous stranger, presumbly trigger fear related brain areas androstadienone might specifically enhance the avoidance tendency towards potentially threatening stimuli, like angry faces.

These hypotheses were evaluated with the approach-avoidance task (Solarz, 1960), a well established method to implicitly assess the connection between higher cognitive functions, motor reactions and motivational tendencies. Participants react to positive or negative stimuli by pulling a joystick towards them or pushing it away. These arm movements reflect approach and avoidance tendencies. Through lifelong conditioning arm flexion is stronger associated with the offset of negative stimuli or the consumption of a desired food than arm extension; in contrast, arm extension is stronger associated with the onset of an aversive stimulus than arm flexion (Cacioppo, Priester, & Berntson, 1993). As a consequence, arm flexion towards positively evaluated objects, which indicates approach behavior, is performed faster than arm extension towards positive evaluated objects, which indicates avoidance behavior. In other words, affect-congruent movements are performed faster than affect-incongruent movements (Chen & Bargh, 1999; Solarz, 1960). And indeed, Rotteveel and Phaf (2004) showed similar results for emotional faces: participants pressed a target button faster when they had to extend their arms to reach the button in response to angry faces, i.e. affect congruency, than when they had to flex their arms to push the button, i.e. affect incongruency. In line with these results faster push than pull reactions to angry faces were reported with a joystick task (Marsh, Ambady, & Kleck, 2005).

Thus, with this method we tested whether and how androstadienone modulates participant´s encoding of and approach and avoidance reactions to emotional faces. We hypothesized that androstadienone would fasten the reaction speed towards angry faces.

Especially the push reaction, reflecting avoidance tendencies, in response to angry faces might be accelerated.

4.1.2 Attitudes and behavioral tendencies

It has been reported that androstadienone affects subjective evaluation of other persons. Saxton and colleagues (2008) reported that women rate male interactions partners as more attractive, when they were exposed to androstadienone compared to sessions in which they were exposed to a common control odor. This implicates that androstadienone affects participant´s attitudes towards other people in a positive way. Attitudes are defined as acquired tendencies to behave in certain ways towards the evaluated object (Campbell, 1963). Attitudes arise through evaluation of the environment on a positive-negative scale, which is necessary to initiate approach or avoidance behavior. Therefore, the resulting behavior seems to be mainly dependent on the valence of the object. Former experiments have shown that specific attitudes towards positive or negative stimuli immediately result in behavioral predispositions towards the stimulus, in such that positive evaluations produce approach tendencies, whereas negative evaluations produce avoidance tendencies (Solarz, 1960). As mentioned above, arm movements seem to reflect approach and avoidance tendencies and also their underlying attitudes (Cacioppo et al., 1993). Affect-congruent movements, i.e. approach behavior towards positive evaluated stimuli, are performed faster than affect-incongruent movements, i.e. avoidance behavior towards positive evaluated stimuli (Chen & Bargh, 1999; Marsh et al., 2005; Rotteveel & Phaf, 2004; Solarz, 1960). These findings indicate a link between negative attitudes and avoidance tendencies, and positive attitudes and approach tendencies. The approach-avoidance task used by these studies is especially appropriate to implicitly assess attitudes and connected behavioral tendencies, because distortion through social desirability is less likely. With this method it is the principle to determine how participant´s reaction speed is influenced by the compatibility between the stimulus valences, the subjective evaluation of those stimuli and the behavioral response.

To sum up, former findings suggest a link between positive attitudes and approach tendencies. Thus with this method it was tested whether and how androstadienone modulates participant´s attitude and following approach and avoidance tendencies towards emotional faces. The hypothesis was that androstadienone would enhance the approach tendency towards faces.

4.2 Methods

4.2.1 Participants

Sixty-two participants (30 women), 18 to 35 years old (M = 24.85; SD = 4.39), with no self reported history of any respiratory, physiological or psychological disease volunteered. All participants had a functional sense of smell which was verified by a 40 items olfactory identification test (MONEX-40) (Albrecht *et al.*, submitted). Two participants, one man and one woman, were excluded due to a self declared status as smokers leaving a total of 60 individuals (29 women) for the final analyses. All participants defined their sexual orientation as exclusively heterosexual according to the Kinsey scale (Kinsey, Pomeroy, Martin, & Gebhard, 1953). All women reported themselves as not being pregnant, reported a regular menstrual cycle and did not use oral contraceptives during the last six months prior testing. Thirty-nine participants defined their ethnical background as White/Caucasian, eleven as Asian, seven as Black/African American and five declined to answer. Women's menstrual cycle phases were determined based on self-reported date of menstrual onset (Jacob et al., 2002). The mean length was 28.7 ±2.0 days, 14% were in their menstrual phase (day 1 – day 6 of the menstrual cycle), 31% in the follicular phase (day 6 – day 14 of the menstrual cycle) and 55% in their luteal phase (day 14 – day 28 of the menstrual cycle). Although defining menstrual cycle phases relative to self-reported menses onset is standard clinical practice, we fully recognize that this referent alone is not as precise as also measuring the preovulatory LH surge and thereby ovulation (Bullivant *et al.*, 2004). However, recently this study's criterion was validated with a larger sample of menstrual cycles (N = 300) with both self-reported menses onset and the day of the preovulatory LH surge documented by hormone assay (Lundström, McClintock, & Olsson, 2006). Collected data indicated that the self-reported criterion used in the study had a very high probability of 98.6% to accurately assign women to the fertile phase. Participants were payed 20 USD for compensation.

4.2.2 Compounds and exposure

The experimental solution consisted of a 250 μM concentration of androstadienone (Steraloid Inc., London; purity > 98%) diluted in propylene glycol (Sigma Aldrich; purity > 99%) with an odor mask of 1 % eugenol (Sigma Aldrich, purity > 99%) to ensure comparability with former studies (Jacob et al., 2002; Jacob, Hayreh et al., 2001; Jacob, Kinnunen et al., 2001; Jacob & McClintock, 2000; Lundström et al., 2003; Lundström & Olsson, 2005; Olsson, Lundström, Diamantopoulou, & Esteves, 2006; Saxton et al., 2008; Villemure & Bushnell, 2007). The control solution consisted of propylene glycol with an odor mask of 1% eugenol. Participants were exposed to the two odors via a constant air flow (2 l/m) using a custom built olfactometer (Boesveldt, Frasnelli, Gordon, & Lundström, 2010).

4.2.3 Material and apparatus

Response times towards emotional faces were measured. To eliminate ambiguous facial features and to isolate the emotion in question, schematic faces with either a happy or an angry facial expression were presented as stimuli (see Figure 1), with the angry face being defined as fear relevant and the happy face as non-fear relevant based on previous studies (Öhman et al., 2001). Stimuli were displayed with *Presentation* software (Neurobehavioral Systems Inc., Albany, CA, USA) on a 19-inch computer screen (resolution: 1280 x 1024 pixels) one meter in front of the participants with a size of 739 x 739 pixels against a black background. Each face was presented 80 times in a randomized order allowing no more than two repetitions of the same facial expression in a row, resulting in 160 trials in total.

Reaction times in response to these cartoon faces was measured with the approach avoidance task (Solarz, 1960), while smelling either androstadienone or the control solution. A joystick was positioned in between the computer screen and the participant. Each trial was initiated by the participant: as soon as the participant pressed the "start" button located near the top of the joystick, the next picture appeared. The participant then had to decide which emotion was expressed by the face and then to pull or to push the joystick accordingly as fast as possible. Faces were shrunken as soon as the lever was moved away from the participant or were enlarged as soon as the lever was moved towards the participant and disappeared finally. "Correct" directional response of the joystick movement to each emotional expression was counterbalanced in two blocks per session. In other words, in one block participants were instructed to pull the joystick towards them in response to a happy face and push the joystick away from them in response to an angry face. Conversely, in the second block they were instructed to respond in the opposite direction. Eight training trials were initiated before each block to allow participants to learn the "correct" direction. We opted for shifting the direction after half of the stimuli to prevent a general effect of direction and to prevent an automatization of participants' responses. To facilitate learning participants got feedback by the word "mistake" on the screen if the joystick was moved in the incorrect direction. Joystick motions to the left or right did not cause any feedback. The time from appearance of the face to the joystick's movement of more than 5° was automatically recorded by the computer. The next picture appeared after pressing the start button again. In each block participants had to react to 40 happy and 40 angry faces. The order of task instructions was for one participant the same in both sessions but randomized between subjects. The response task lasted about 10 minutes including instructions.

Figure 1. The schematic angry and happy facial expressions used in the study

4.2.4 Procedure

A counterbalanced within subjects design was used, for which each participant underwent two testing sessions on two separate but not consecutive days. Both sessions had to be conducted within four days. Each session took about 30 minutes and followed an identical protocol with applying either the control or the experimental solution. Sequence of odors was randomized between subjects. To control for circadian changes in hormone levels and alertness, all participants had their two testing sessions at the same time of day. A female experimenter completed all interactions with the participants, who were blind to the applied odor solution and was present in the testing room at all time. After filling out written informed consent (see 8.1.1), a cannula connected to the olfactometer was fitted to the participant for odor application. Initially, participants rated the odors, either androstadienone (in the experimental session) or the control substance (in the control session), for intensity, pleasantness and familiarity on 100 mm visual analog scales (see 8.4). They subsequently followed onscreen instructions (see 8.1.2) with minimum experimenter interaction and performed the response task. At the end of the task, participants rated the happy and angry face for perceived intensity and pleasantness on 100mm visual analog scales (see 8.4) while smelling the applied odor. At the very end, the participant's ability to identify odors was assessed with the MONEX-40 (Albrecht et al., submitted).

4.2.5 Data analyses

Only correct responses and reaction times above 100ms were analyzed. Error rates (i.e. incorrect responses) were very low and averaged less than 2.6 % with no significant variation between odor sessions, experimental conditions or participant's gender. Responses considered as outliers (±3 SD) were identified and removed separately for each individual, instruction and odor. Average percentage of outliers did not exceed 5%, i.e. two outliers per 40 reactions, per individual, instruction or odor. Mean reaction times for each individual, response direction, emotion and odor were calculated.

4.2.5.1 Attention-based reactions

Statistical comparisons were assessed using repeated-measurement ANOVAs with direction (push vs. pull), emotion (angry vs. happy) and odor (androstadienone vs. control) as within factors and participants' sex as between subjects factor. If necessary, Greenhouse-Geisser corrections of degrees of freedom were applied for violations of sphericity. Interactions were followed by *Student's t*-tests.

4.2.5.2 Attitudes and behavioral tendencies

Several studies assessing attitudes and consequential behavioral responses calculated the approach-avoidance score, an index, which reflects specific action tendencies (Greenwald, McGhee, & Schwartz, 1998; Heuer, Rinck, & Becker, 2007; Neumann, Hulsenbeck, & Seibt, 2004; Rinck & Becker, 2007). The approach-avoidance score is determined by subtracting each participant's mean reaction time in the pull condition from the mean reaction time in the corresponding push condition (e.g. angry-push minus angry-pull, happy-push minus happy-pull), which reflects the relative direction of the response tendency: positive values indicate stronger approach than avoidance tendency and negative values indicate stronger avoidance than approach tendency. Repeated-measurement ANOVAs with emotion (angry vs. happy) and odor (androstadienone vs. control) as within factors and participant's sex as between subjects factor were calculated. If necessary, Greenhouse-Geisser corrections of degrees of freedom were applied for violations of sphericity. Interactions were followed by *Student's t*-tests.

4.3 Results

4.3.1 Odor ratings

Subjective ratings of androstadienone and the control solution revealed similar results (see Table 1). ANOVA with odor and participant's sex separately for intensity, pleasantness and familiarity did not reveal any significant effects (all $ps > .38$), indicating no conscious discrimination between the control and the experimental odor solution and no differences in sensitivity between men and women.

Table 1. Subjective intensity, pleasantness and familiarity ratings of androstadienone and the control solution (SD in brackets).

	Androstadienone	Control
Intensity	20.8 (13.5)	21.9 (16.6)
Pleasantness	59.0 (13.8)	59.3 (15.0)
Familiarity	45.8 (23.6)	42.7 (24.8)

4.3.2 Face ratings

Exposure to androstadienone did not alter face ratings in a significant way (see

Table 2). ANOVA demonstrated that the angry face was perceived as more intense than the happy face (main effect of emotion: $F(1, 58) = 21.08$, $p < .001$, $\eta^2 = .27$). The happy face was rated as significantly more pleasant than the angry face (main effect of emotion: $F(1, 58) = 526.56$, $p < .001$, $\eta^2 = .90$). Women rated the facial expressions as less intense than men (46.30 vs. 57.59, main effect of participant's sex: $F(1, 58) = 5.36$, $p = .024$, $\eta^2 = .09$).

Table 2. Intensity and pleasantness ratings of angry and happy cartoon faces, while inhaling androstadienone or the control solution (SD in brackets)

	Androstadienone		Control	
	Angry	Happy	Angry	Happy
Intensity	56.9 (22.5)	47.5 (20.8)	56.7 (22.3)	47.4 (22.7)
Pleasantness	16.3 (14.2)	77.2 (11.7)	19.4 (19.0)	75.8 (12.7)

4.3.3 Attention-based reactions

Overall, the pull reaction was significantly faster than the push reaction (628 ms vs. 641 ms), $F(1, 58) = 18.24$, $p < .001$, $\eta^2 = .24$. This effect was not significantly modulated by androstadienone (Odor x Direction, $p > .19$).

Reactions to angry faces were significantly faster than reactions to happy faces (631 ms vs. 638 ms), $F(1, 58) = 8.30$, $p = .006$, $\eta^2 = .13$. This effect was modulated androstadienone exposure: androstadienone significantly accelerated reaction speed towards angry faces, independently of participant's sex (Odor x Emotion: $F(1, 58) = 4.63$, $p = .036$, $\eta^2 = .07$; Odor x Emotion x Participant's sex interaction: $p > .87$). As hypothesized, participants reacted significantly faster to angry faces when they were exposed to androstadienone compared to the control solution (627 ms vs. 637 ms), $t(59) = 2.33$, $p = .023$. In contrast, reaction speed to happy faces was not modified by odors (638 ms vs. 638 ms, $p > .89$) (see Figure 2).

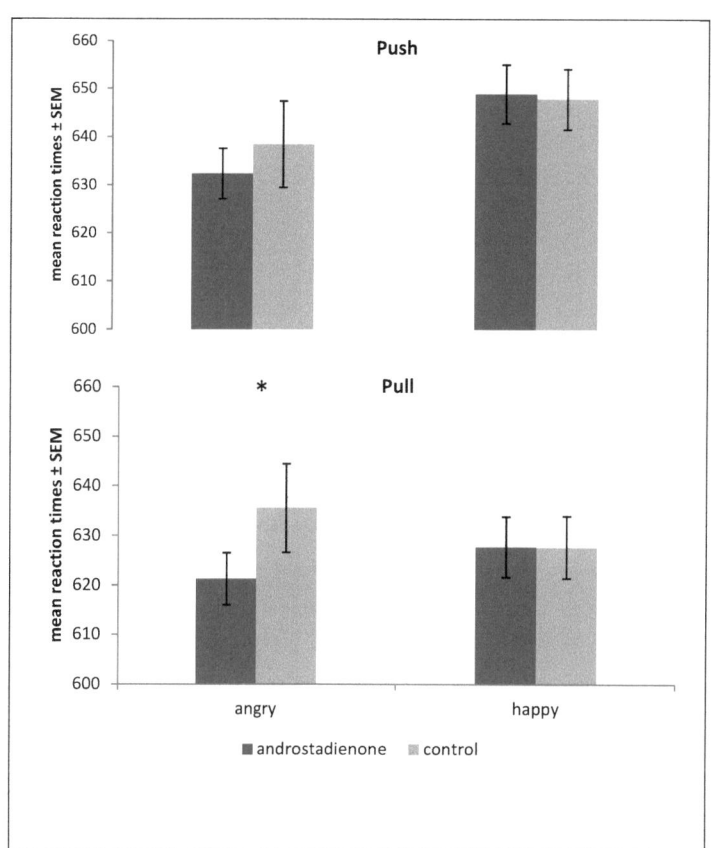

Figure 2. Mean reaction times in milliseconds for each odor and facial expression combination for both the push and the pull direction. Asterisk indicates the significant odor effect at $p < .05$ in response to angry faces.

4.3.4 Attitudes and behavioral tendencies

ANOVA revealed a significant interaction between odor and participant´s gender, $F(1, 58) = 4.31$, $p = .043$, $η^2 = .07$. Separate t-tests comparing odors in men and women revealed for men significance, $t(30) = 2.69$, $p = .011$. t-tests testing the difference of the approach-avoidance score from zero reached significance for men in the experimental condition, $t(30) = 3.89$, $p < .001$, indicating stronger approach tendency, whereas in the control condition men showed neither approach, nor avoidance tendencies (all $ps > .23$) (see Figure 3). For women the approach-avoidance score was significantly different from zero in both, the experimental, $t(28) = 2.11$, $p = .044$ and the control condition, $t(28) = 3.36$, $p = .002$, indicating a general approach tendency in women towards emotional faces.

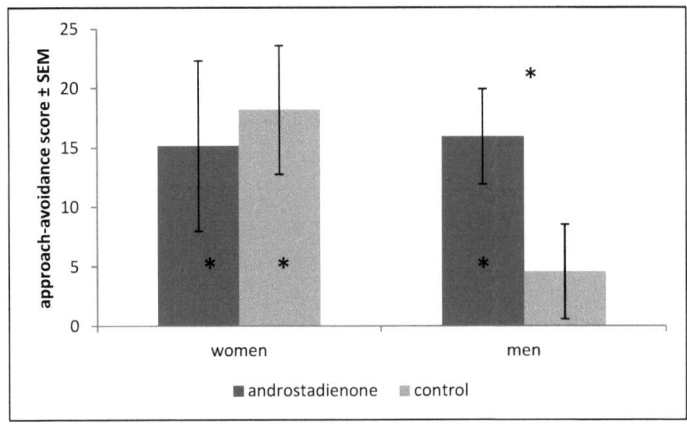

Figure 3. Approach-avoidance score (±SEM) for each odor and participant's sex combination. Asterisks indicate significant differences of the approach-avoidance score from zero and between odors in men at $p < .05$.

4.4 Discussion

4.4.1 Attention-based reactions

The results of the current study indicate that the endogenous compound androstadienone enhances the processing of visual threatening stimuli. As suggested, arm movements were accelerated in response to angry faces when participants were exposed to androstadienone compared to a control odor. Notably, this was independent of the claimed direction of arm movement, suggesting that both directions were influenced in a similar way. The literature has repeatedly shown that anger is an important social signal which is processed and attended preferentially because of signaling potential threat (Öhman et al., 2001; Wieser, Pauli, Reicherts, & Mühlberger, 2010). Produced by our conspecifics, androstadienone signals the presence of a possible interaction partner in the immediate vicinity, who might at the same time be potentially threatening. Consequently, the negative olfactory priming seems to enhance the evolutionary primed attention towards threatening visual stimuli, resulting in a facilitated motor reaction. This is in line with former research suggesting cross-modality priming (Pauli, Bourne, Diekmann, & Birbaumer, 1999).

That androstadienone enhances attention has already been suggested by several authors. A better self reported feeling of paying attention or being more focused sustained by androstadienone (Hummer & McClintock, 2009; Lundström et al., 2003) as well as an enhanced attention towards a negative stimulus with androstadienone exposure (Villemure & Bushnell, 2007) were reported. Also, Hummer and McClintock (2009) showed that androstadienone enhances attention specifically towards emotional compared to neutral faces in a dot probe task; however, contrary to our results they did not find emotion specific differences. The dot probe task measured attention implicitly. Emotional faces were presented subliminal so that participants were unaware of the displayed facial expressions, and focused on detecting the following dot appearing on the screen. In the present study, participants were explicitly instructed to identify the expressed emotion in order to perform the response task thus triggering a more conscious processing of the emotion than participants in Hummer and McClintock's study. Moreover, Hummer and McClintock (2009) used real faces thus introducing a variance in the level of expressed emotion between faces. Schematic faces, as used in this study, are controlled for idiosyncrasy, i.e. the evolutionary significant signal "threat" did not vary between fear related stimuli. Therefore, in contrast to Hummer and McClintock (2009), this negative signal might have gotten specific attentional focus, which in turn might have been enhanced by androstadienone.

However, an interaction between movement directions and emotions, the measurement for approach-avoidance tendencies, did not reach significance. In contrast to previous data (Marsh et al., 2005; Rotteveel & Phaf, 2004) we did not detect a significant faster push than pull reaction in response to an angry expression in the control condition, which should have reflected enhanced avoidance. This discrepancy may be due to different stimulus material. While we used one genderless angry cartoon face, Marsh and colleagues (2005) presented photographs of four different men and women, whereas Rotteveel and Phaf (2004) even used 20 different men and women. This indicates that rather than an isolated emotional expression, additional vital facial features may be important for approach-avoidance reactions. Moreover, Marsh and colleagues (2005) compared angry with fearful faces, but not with happy faces, as in the current study, which also may interfere with reaction tendencies. Whether this combination may influence brain processing and motor reactions warrant further investigations in future studies.

Although expected, androstadienone did not enhance the avoidance reactions to the angry face specifically, which should have been reflected by a faster push than pull movement. Besides above mentioned methodological reasons, this may also suggest that the endogenous odor has no differential effect on our motivational approach – avoidance systems. To initiate approach or avoidance behavior evaluation of the environment on a positive-negative scale is necessary.

Thus, the resulting behavior seems to be mainly dependent on the valence of the object. However, supported by our valence rating data, androstadienone seems not be able to modulate the evaluation of the emotional cartoon faces. In other words, rather than the valence intensity or pleasantness, the speed of exact threat decoding might have been fastened by androstadienone. This is supported by our accuracy data, where correct response rates in the experimental condition were similar to that in the control condition, as well previous published data demonstrating that androstadienone received preferential processing of the human brain (Lundström, Olsson et al., 2006).

Like other studies (Bensafi, Brown et al., 2004; Hummer & McClintock, 2009; Jacob et al., 2002), we did not find any sex specific androstadienone modulations. However, some former studies reported a sex specificity of androstadienone effects in altering brain functions or physiological and psychological reactions (Bensafi, Brown et al., 2004; Jacob, Kinnunen et al., 2001; Jacob & McClintock, 2000; Savic et al., 2005). Methodological differences between study designs might be responsible. First, with onscreen instructions we kept the social interaction, a common explanation for sex-specific effects, between experimenter and subjects at a minimum. Second, Bensafi et al. (2004) showed sex-specificity in mood effects only in an emotional arousing context induced by sad film clips. If, however, Bensafi and colleagues (2004) tested men and women in a non-arousing context, as we did in the current study, men´s and women´s physiological reactions to androstadienone exposure were comparable. Therefore, sex-specific androstadienone effects might have their origin in the combination of contextual and social circumstances as well as individually different characteristics, like e.g. sexual orientation (Berglund et al., 2008; Berglund, Lindström, & Savic, 2006; Savic et al., 2005).

To conclude, androstadienone, undetectable as an odor, is able to alter behavior, which was demonstrated by accelerated arm movements in response to a schematic angry face. This suggests that androstadienone modulates how the mind reacts to negative visual information. Androstadienone, communicating a potentially threatening social interaction approach, might allocate attentional resources specifically towards threatening social stimuli, and with it, enhance the preparation for appropriate actions. These findings indicate that a none-conscious concentration of the endogenous odor androstadienone modulate ongoing behavioral responses. Although the current study provides further support of the biological relevance of androstadienone, a replication of these effects in ecological valid everyday life situations using ecological concentrations of the active compound would strengthen this assumption. Nonetheless, current results corroborate the notion of androstadienone as an active chemosignal potentially modulating attentional processes, which in turn results in changed behavioral reactions.

4.4.2 Attitudes and behavioral tendencies

Results given by the approach-avoidance score analyses suggest that androstadienone enhances men's approach tendencies towards faces independently of emotional expressions. In contrast, women's behavioral tendencies were not affected by androstadienone. According to our hypothesis, the endogenous odor might have strengthened positive attitudes towards faces in men and therefore facilitated their approach tendencies towards them. Rotteveel and Phaf (2001) suggested that a conscious appraisal procedure is necessary to elicit specific behavioral tendencies, at least concerning arm flexion and extension movements. In the current study, as in Rotteveel´s and Phaf´s (2001), participants were explicitly instructed to attend to affective features, namely the expressed emotion. Therefore, it seems plausible that androstadienone, rather than acting on automatic reflexive processes, might modulate higher cognitive functions which are involved in processes transferring appraisal into motoric responses. Thus, according to our hypothesis, androstadienone might have improved men´s evaluations leading to an enhanced willingness to approach. Consequently, as a social odor, androstadienone might serve a pro-social function in men.

Based on these findings one would expect that subjective face ratings of pleasantness were improved by androstadienone. However, androstadienone did not significantly modulate explicit evaluation, indicating an implicit effect mechanism of androstadienone. This is in line with a former study which also did not detect any androstadienone effects on explicit attractiveness ratings (Lundström & Olsson, 2005). Why then Saxton and colleagues (2008) found androstadienone related evaluation effects? The most striking difference between studies is the context. Saxton et al. (2008) conducted the study in a complex, lively, social situation, contrasting our and Lundström `s and Olsson`s (2005) experiment, which were conducted in a quiet laboratory room without any distraction. This indicates a high significance of appropriate situations to assess androstadienone effects on explicit subjective evaluation as has been argued recurrently (e.g. Hummer & McClintock, 2009).

Moreover, the direct influence of androstadienone on subjective attitudes is still challenged. Although former studies found significant connections between implicit attitudes and behavioral tendencies

(Neumann et al., 2004), we cannot certainly conclude from an enhance approach tendency by androstadienone exposure a positively changed appraisal and attitude, because no additional measures of attitudes were assessed. Taking into account that androstadienone is produced by both men and women it certainly communicates the presence of an individual, but not the gender of the encounter. Men´s reaction might depend on being confronted with a male or female person, who is either a friend or a rival. Assuming that a male individual signals a potential competitor for men, the enhanced approach tendency with androstadienone might reflect aggression rather than positive intentions. Rather pro-social responses may occur if the intruder is a familiar or friendly person or a woman, who elicits sympathy, the desire to help or sexual interest. In other words, androstadienone might strengthen attitudes dependent on the intent of a specific situation: androstadienone may enhance an aggressive approach to a male, but a positive approach to a potential mate partner. Future studies using male and female faces as well as strangers or familiar persons might be able to clarify this issue.

Rather unexpected was the result that women showed already in the control condition an enhanced approach tendency, which in turn was not affected by androstadienone. Moreover, already in the control session men and women differed significantly in their behavioral predispositions; women showed higher approach tendency to faces than men. An evolutionary perspective suggests that women are highly in need for support by other group members, because of their longterm responsibilities for their offspring. Therefore, women´s basic attitude towards conspecifics might be more positive and in turn result into facilitated approach behavior. Androstadienone then might not be able modulate or, even less, attenuate this ecological meaningful mechanism.

It rather may maintain the enhanced approach tendency in women. However, this explanation remains speculative, because to our knowledge no other study has explored gender differences with respect to approach-avoidance tendencies towards emotional faces or other affective stimuli.

Nevertheless, current findings provide support that androstadienone is an effective substance modulating higher cognitive mechanisms, which in turn results in changed behavioral tendencies towards face stimuli.

5 Cortical reactions

5.1 Introduction

Androstadienone modulates brain activity in areas associated with emotional and attentional processes and social cognition (Gulyas et al., 2004; Jacob, Kinnunen et al., 2001). Activity changes in areas like the occipital and the parietal cortex suggested an influence of androstadienone on attentional processes especially regarding visual stimulation (Jacob, Kinnunen et al., 2001). An interesting question is, whether androstadienone might indeed affect the central nervous perception of visual stimuli. This can be determined by measuring event related brain potentials (ERPs). Different components of visual ERPs enable us to track attentional processes in the brain in milliseconds resolution and to discern between relevant, i.e. attended, and dispensable, i.e. unattended, stimuli (Schupp, Junghöfer, Weike, & Hamm, 2003a). As an early positive peak occurring around 100 ms after stimulus onset, the P100 reflects early automatic processing within the visual cortex and responds to manipulations of selective attention, with more attended stimuli eliciting larger amplitudes (Luck, Woodman, & Vogel, 2000). A valence rather than arousal dependence of the visual P100 was assumed by several studies because of larger amplitudes in response to unpleasant compared to pleasant valence categories with matched arousal levels (c.f. Olofsson, Nordin, Sequeira, & Polich, 2008). Furthermore, a later positive shift starting around 300 ms after stimulus onset, the P300 reflects neural activity related to cognitive processes like attention allocation and updating of the working memory (Polich & Kok, 1995). If the current stimulus is novel or meaningful, attention processes are engaged and reflected by an enhanced P300. This component is influenced by motivation, task relevance and arousal level. Two further

ERP components are especially suggested to reflect the processing of the pictures' affective significance: the early posterior negativity (EPN) and the late positive potential (LPP). An EPN occurs as a negative deflection to emotional as compared to neutral stimuli about 240 ms after picture onset. This potential can be observed in response to positive and negative scenes (Schupp et al., 2003a) and facial expressions like anger (Schupp et al., 2004), fear and happiness (Mühlberger et al., 2009; Sato, Kochiyama, Yoshikawa, & Matsumura, 2001) compared to a neutral picture or facial expression. The EPN is assumed to reflect facilitated perceptual processing at a stimulus driven preconscious level, indexing natural selective attention in such that affectively arousing stimuli are tagged for further processing. The LPP is a latter portion of the ERP waveform occurring around 400 ms post stimulus over a broad latency interval as a slow positive wave. It arises in response to high arousing emotional pictures and facial expressions, as compared to neutral or less attended stimuli reflecting a top down regulation of more conscious perceptual processing, which is also modulated by the intrinsic relevance of presented stimuli (Lang, 1997; Schupp et al., 2003a). Both, the EPN and LPP depend on the stimulus arousal level since highly arousing pictures like mutilations or erotica elicit stronger EPN and LPP amplitudes than less arousing pictures (Cuthbert, Schupp, Bradley, Birbaumer, & Lang, 2000; Schupp et al., 2003a). In sum, affective stimuli are preferentially processed in the brain. Moreover, the processing of facial features is indicated by a negative peak, N170, around 170 ms post stimulus bilateral on occipito-temporal electrodes (Bentin, Allison, Puce, Perez, & McCarthy, 1996; Eimer & McCarthy, 1999; Linkenkaer-Hansen et al., 1998). This facial peak, interpreted as a measure of the structural encoding of faces, occurs in response to photographs, paintings as well as schematic faces (Mühlberger et al., 2009; Sagiv & Bentin, 2001).

Whether androstadienone has the capacity to modulate the cortical processing of visual stimuli was explored in two different experiments. In study II, androstadienone related modulations of the central nervous processing of social and non-social visual stimuli were investigated in women. In study III, androstadienone effects on face processing were tested in men. Both studies sought for clarification of mechanisms through which androstadienone may be operating.

5.2 Study II: cortical reactions in women

This study investigated influences of androstadienone on women´s visual cortical processing. Specifically, four different categories of visual stimuli with positive, negative and neutral valence were presented to clarify the type of stimuli, i.e. the specific content of pictures, androstadienone is acting on. First, faces with different emotional expressions simulating an immediate interaction partner by directly facing the observer. Second, affective scenes with social content, i.e. pictures with groups of people, were presented to simulate a rather passive observing condition of potential interaction partners. Third, scenes with heterosexual couples were used to simulate a sexual context to the observer. Fourth, scenes without any persons served as the non social control category.

In general, attended or important compared to dispensable stimuli should elicit larger amplitudes evident in the P100 and P300 component. The EPN and LPP amplitudes reflect the enhanced processing of affective or attention capturing stimuli compared to emotionally neutral pictures. On the other hand, androstadienone is suggested to enhance attention in general but also specifically towards emotional information.

Thus, we expected larger P100 or P300 amplitudes in reaction to presented visual stimuli, but also larger EPN and LPP amplitudes in response to emotional related pictures, while smelling androstadienone compared to a control odor. Moreover, we hypothesized that androstadienone as a social odor might rather affect the processing of social compared to non-social stimuli. To specifically address the effect of androstadienone on faces, we also examined the face specific negative peak N170. To provide a behavioral correlate of attentional processes participants had to conduct two alertness tasks measuring sustained attention and the rapid recruitment of awareness. We expected better performance in behavioral attention tasks if participants smell androstadienone. To replicate earlier findings and to provide control variables we measured androstadienone effects on physiology and mood.

5.2.1 Methods

5.2.1.1 Subjects

Fifty-one non smoking women (23 in control group, 28 in androstadienone group), without nasal congestion and between 18 and 38 years (groups matched for age: control group: $M = 24.16$, $SD = 4.26$; androstadienone group: $M = 23.78$, $SD = 2.99$) participated. All women reported to take hormonal contraceptives for at least six months prior to the testing session. The participants were students of the University of Würzburg and recruited by an online advertisement. All described themselves as heterosexual and right-handed and had no respiratory or psychological disease. Participants gave their written informed consent, conforming to institutional guidelines for human research (approved by the local Ethic committee) and were paid 15 Euro for compensation.

5.2.1.2 Compounds

Our experimental solution consisted of a 250 µM concentration of androstadienone (Steraloid Inc., London; purity > 98%) in propylene glycol (Sigma Aldrich; purity > 99%) with an odor mask consisting of 1 % eugenol (Sigma Aldrich, purity > 99%) to ensure comparability with former pheromone studies (Jacob, Hayreh et al., 2001; Jacob & McClintock, 2000; Lundström & Olsson, 2005; Olsson et al., 2006; Saxton et al., 2008; Villemure & Bushnell, 2007). The control solution consisted of propylene glycol with an odor mask of 1% eugenol.

5.2.1.3 Psychological measurements

Mood and anxiety. In order to replicate mood effects of androstadienone (Jacob & McClintock, 2000; Lundström & Olsson, 2005), the same eight adjectives applicable to current mood as in previous studies were administered: *social, open, relaxed, heavy, focused, sensual energetic, irritated* (German translation: *sozial, offen, entspannt, schwermütig, fokussiert, sinnlich, aktiv, iritiert*); subjects rated their response by marking a 100 mm Visual Analog Scale (VAS; Mackay, 1980), ranging from 0, not at all, to 100, extremely. Additionally, the German version (Laux, Glanzmann, Schaffner, & Spielberger, 1981) of a standardized psychometric anxiety questionnaire was used: the State Trait Anxiety Inventory-State (STAI-S; Spielberger, Gorsuch, & Lushene, 1970). To control for mood and anxiety changes over time, participants rated their actual mood and anxiety nine times during the experiment. Measurements before exposure to the solution were subtracted from the measurements after the exposures for all VAS and STAI-S scales to reflect the potential time effects of androstadienone.

Alertness. Phasic and tonic alertness was assessed by measuring reaction speed to a suddenly appearing visual cue with and without a preceding warning stimulus (Sturm & Willmes, 2001; Tales, Muir, Bayer, Jones, & Snowden, 2002). Phasic alertness is the rapid mobilization of resources to process an expected stimulus. It is also called extrinsic alertness, because the attention allocation receives an external support by a visual warning signal preceding the cue. In other words, it represents the ability to increase response readiness subsequent to an external cue. Tonic alertness, also called intrinsic alertness, is the ability to internally sustain general attention over time. This was tested to an incidentally appearing cross on the computer screen. Subjects had to react as fast as possible to the cross by pressing a key on the keyboard. Inter stimulus intervals (ISIs) varied between 1 and 1.5 seconds. The target remained on the screen until a response was made. Before starting the task five training trials were presented to familiarize the participants with the procedure. The test consists of two blocks with ten trials each separated by a 15 seconds break. In the second condition the phasic alertness was tested. This time a warning stimulus preceded the cross, to which the participants had to react again as fast as possible. The warning stimulus appeared for durations of 100 ms, 400 to 700 ms prior to the cross. This task also consists of five training trails and two ten trial blocks separated by a 15 seconds break. The target remained on the screen until participant responded.

5.2.1.4 Physiological measurements

To control for physiological effects of androstadienone and to replicate previous results (Bensafi, Brown et al., 2004; Bensafi et al., 2003) autonomous nervous system (ANS) reactions like cortisol release, galvanic skin response (GSR), heart rate (HR) and respiration rate (RR) were measured. Using V-Amp (Edition Version: 1.03; *Brain Products,*

Munich, Germany) and Brain Vision Recorder (*Brain Products,* Munich, Germany) GSR, HR and RR were recorded with a 500 Hz sampling rate.

Cortisol. Saliva was collected in SaliCap sampling tubes (*IBL*, Hamburg) three times during the experiment, once before odor application as baseline measurement and two times after odor application; 20 minutes after the first and 20 minutes after the second application. Samples were stored at -20°C. Saliva cortisol levels were analyzed by Dresden LabService GmbH.

Skin conductance response (EDA). Skin conductance in micro Siemens (µS) was recorded by two conductive gel filled Ag/AgCl electrodes (diameter 1 cm) placed on the inner side of the left hand. Two subjects were excluded because of recording problems, resulting in 23 controls and 26 subjects in the androstadienone group.

Heart rate (HR). HR was recorded by means of two Ag/AgCl electrodes places at the sternum and the abdomen and calculated from the R-wave of the electrocardiogram (ECG). Data were reduced to beats per minute (BPM). Five subjects were excluded because of recording problems, resulting in 22 control subjects and 24 in the androstadienone group.

Respiration rate. Respiration was measured by a respiratory belt transducer measuring changes in thoracic circumference due to respiration with a sensitivity of 1 mV/mm. The data were reduced to respiration rate per minute. Two subjects were excluded because of recording problems, resulting in 23 subjects in the control group and 26 subjects in the androstadienone group.

5.2.1.5 Stimulus material and presentation

Sixteen photographs of human faces (8 men, 8 women) from the Karolinska Directed Emotional Faces (KDEF, Lundquist, Flykt, & Öhman,

1998) battery were used, each with a happy, angry and neutral facial expression[1]. Stimuli were displayed with *Presentation* software (*Neurobehavioural Systems Inc.*, Albany, CA, USA) on a 19-inch computer screen (resolution: 1024 x 768 pixels) one meter in front of the participants with a size of 562 x 762 pixels against a grey background. Each face was presented twice resulting in 32 trials for each emotion and 96 faces in total. Randomization was restricted in order to allow not more than two repetitions of the same facial expression (angry, happy or neutral). In a second block, 45 affective scenes[2] from the International Affective Picture System (IAPS; Lang, Bradley, & Cuthbert, 1999) were presented. They were grouped into three different categories: category "couples" contained scenes showing heterosexual couples, category "social" contained social scenes showing one or more people and category "non-social" contained non-social scenes not showing any person. Each category was then separated for valence, resulting in five negative (e.g. violence, mutilations, trash), five positive (e.g. erotic scenes, laughing people, flowers) and five neutral pictures (e.g. a promenading couple, talking people, a book). Each picture was shown six times, resulting in 270 trials in total. A fixation cross displayed for 500 ms preceded each stimulus to ensure that participants were looking at the centre of the screen. Each stimulus was presented for 1000 ms with a grey screen ISI varying between 1.5 and 2.5 seconds. The total presentation of faces and scenes lasted about 25 minutes.

---

[1] **KDEF stimuli:** AF03, AF05, AF07, AF14, AF22, AF26, AF27, AF34, AM01, AM10, AM11, AM12, AM18, AM26, AM29, AM35
[2] **IAPS:** positive couples: 4660, 4680, 4670, 4650, 4689; negative couples: 6312, 6315, 6360, 6530, 6561; neutral couples: 4605, 4598, 4625, 4606, 4609; positive social: 5621, 8370, 8380, 8461, 8497; negative social: 2691, 3500, 3530, 6562, 6821; neutral social: 2222, 2579, 2590, 2593, 2595; positive non-social: 5811, 7330, 8170, 8502, 5260; negative non-social: 1120, 7380, 9290, 9301, 9140; neutral non-social: 5510, 7002, 7025, 7090, 7500;

5.2.1.6 Event related brain potential measurement and analyses

The electroencephalogram (EEG) was continuously recorded with a sampling rate of 1000 Hz from twenty-eight electrodes mounted on a flexible cab according to the international 10/20 system (EASYCAP GmbH; Fp1, Fpz, Fp2, F7, F3, Fz, F4, F8, T7, C3, Cz, C4, T8, P9, P7, P3, Pz, P4, P8, P10, PO7, O1, O2, PO8, PO9, O9, O10, PO10). An electrode at Fcz was used as ground electrode and the right mastoid (M2) as implicit reference. The impedance for each Ag-AgCl-electrode was kept below 5 kΩ and amplifier band pass was set to 0.1 to 100 Hz online, using a Brain-Amp-MR amplifier (*Brain Products*, Munich, Germany) and the software *Brain Vision Recorder* Version 1.04 (*Brain Products*, Munich, Germany). Vertical (electrodes above and below left eye) and horizontal (electrodes at outer canthi of both eyes) eye movements were recorded by electrooculogram. The recorded datasets were processed off-line with the software Brain Vision Analyzer Version 2.0 (*Brain Products*, Munich, Germany). First, EEG data were low-pass and high-pass filtered (0.1 Hz and 30 Hz, respectively) and afterwards re-referenced to linked mastoids. Then, epochs from 100 ms before until 700 ms after picture onset were extracted and corrected for blink artifacts using a digital ocular correction (Gratton, Coles, & Donchin, 1983). Baseline correction was performed using the 100-ms pre-stimulus interval. Automatic artifact rejection used a maximal allowed voltage step of 50 µV per milliseconds and maximal allowed amplitude of ± 50 µV individually for each channel. Then, epochs were averaged separately for each facial expression and actor's sex, as well as picture category and valence, for each electrode and participant. The P100 peak was searched in a time window between 50 ms and 130 ms post stimulus on

O1 and O2 electrodes, the N170 peak between 130 ms and 200 ms post stimulus on P7 and P8 electrodes. EPNs were estimated as mean activity between 280 ms and 310 ms post stimulus over left hemispherical (PO7, PO9, O1 and O9) and analogous right hemispherical electrodes (PO8, PO10, O2 and O10). P300 amplitudes were calculated as mean activity between 320 ms and 360 ms after stimulus onset on P3, Pz and P4 electrodes. LPPs were estimated as mean amplitude between 400 ms and 600 ms post stimulus over midline electrodes (Fz, Cz, Pz).

5.2.1.7 Procedure

For an overview of the experimental procedure see Table 3. A double-blind between-subjects design was used such that each subject underwent one session. Potential influences of circadian rhythms were minimized by conducting testing sessions at least three hours after the usual wake up time. Both groups were tested on average on the same time of day. Testing was performed individually in a sound attenuated laboratory room. Activity was continuously monitored from the adjacent control room via a video monitor. A same-sex experimenter completed all necessary interactions. To hold them at a minimum all instructions were presented on a computer screen (see 8.2.2). After obtaining written informed consent (see 8.2.1) ECG and EDA electrodes and the respiration belt were attached. During mounting of the EEG equipment, participants answered demographic questionnaires. Then, left alone in the experimental room, baseline ratings of mood and anxiety were obtained and subsequently a three minutes baseline of all physiological measures, heart rate, skin conductance and respiration rate, was recorded. The experimenter re-entered the room, collected the baseline salivary sample and applied then one milliliter of the experimental or control solution on a cotton swap. This was then rubbed onto the skin area between the upper lip and the nostrils as has been done in previous studies (Hummer & McClintock, 2009; Jacob, Hayreh et al., 2001; Jacob & McClintock, 2000; Lundström et al., 2003; Olsson et al., 2006; Saxton et al., 2008). The experimenter told the subject to follow onscreen instructions by using the PC keyboard and left the cabin.

During the first 20 minutes after the first substance application alertness tasks were conducted. In between, subjects rated four times their actual mood and anxiety. Then, the second salivary sample was collected and afterwards the second solution application was done by the participant herself to keep social interaction at a minimum. After the second substance application participants rated their mood and anxiety four more times until the end of the experiment. Next, while subjects passively viewed the face and picture presentation, EEG was continuously recorded. Finally, they rated all presented faces for valence, arousal, attractiveness and sympathy, the IAPS pictures for valence and arousal and the applied odor for pleasantness, intensity and familiarity. Then, the third saliva sample was collected. Afterwards, the experimenter re-entered the cabin and conducted a discrimination task. Subjects had to tell the experimental and control solution apart and had to guess which one they smelled during the experiment. Correct discriminations did not reach above chance level. Then, the experimenter disconnected all recording devices and the subject got paid.

Table 3. Experimental course of study II.

Procedure	Duration [min]	Time [min]
mounting equipment	45	0
mood and anxiety baseline rating	1	46
physiological baseline	5	51
salivary cortisol baseline sample	2	53
1st compound application	1	54
4 mood and anxiety ratings and alertness tasks	20	74
2nd saliva sample	2	76
2nd compound application	1	77
6th mood and anxiety rating	1	78
EEG recording	25	103
7th mood and anxiety rating	1	104
subjective stimulus rating	10	114
8th mood and anxiety rating	1	115
3rd saliva sample	2	117
9th mood and anxiety rating	1	118
discrimination task	2	120

5.2.1.8 Data analyses

Subjective stimulus ratings. For face ratings ANOVA with actor's sex and emotion as within factors and odor as between subjects factor were calculated separately for each, facial valence, arousal, attractiveness and sympathy. For scene ratings ANOVA with category and valence as within and odor as between subjects factor were conducted separately for valence and arousal ratings of scenes.

Subjective mood ratings. Mood and anxiety ratings were expressed as change scores by subtracting the baseline value from each of eight experimental values, four values after the first odor application and four values after the second odor application. This results in eight change scores for each mood and state anxiety variable, reflecting the time course of substance effects. Positive values indicate an odor related increase, while negative values indicate an odor related decrease. Two separate repeated measurement ANOVAs, for each four variables after the first and after the second odor application, were calculated. The within subjects factor was time and the between subjects factor was odor. Due to missing values of one woman in the androstadienone group mood variables of 23 participants in the control and 27 participants in the androstadienone group were analyzed. Due to missing values of two women in the control and six women in the androstadienone group state anxiety variables of 22 participants in the control and 22 participants in the androstadienone group were analyzed.

Alertness. After initial visual inspection data scores below 150 ms and above 1500 ms were eliminated as anticipatory and due to lapses of concentration or misunderstandings of the task performance. Intrinsic alertness was calculated as median of reaction times for each participant. Extrinsic alertness was then calculated by subtracting the median of the tonic alertness task from the median of the phasic alertness task (Tales *et al.*, 2002). *Student´s t*-tests of means of participant´s medians of intrinsic and extrinsic alertness scores were calculated between odor groups.

Physiology. All data were analyzed with Brain Vision Analyzer (*Brain Products*, Munich, Germany). Mean values for GSR, HR and RR were calculated over time intervals of three minutes. The baseline interval before compound application, four intervals 5, 10, 15 and 20 minutes after the first compound application, and four intervals 5, 10, 15 and 20 minutes after the second compound application were calculated, resulting in nine mean values for each physiological measurement. Then, data were expressed as change scores by subtracting the baseline value from each of eight experimental values, resulting in eight scores for each physiological response, reflecting the time course of substance effects. Positive values indicate a substance related increase, while negative values indicate a decrease in physiological arousal in response to applied odors. ANOVAs were calculated separately for the four scores after the first and the four scores after the second compound application, with time as within factor and odor as between subjects factor. Due to recording problems of heart rates of one woman in the control and four women in the androstadienone group, heart rate data of 22 participants in the control and 24 participants in the androstadienone group were analyzed. Due to recording problems of skin conductance and respiration rates of two women in the androstadienone group, skin conductance and respiration rate data of 23 participants in the control and 26 participants in the androstadienone group were analyzed.

Cortisol data were expressed as change scores by subtracting the baseline concentration from each of two experimental concentrations resulting in two change scores. Positive scores reflect an odor related increase in salivary cortisol levels, while negative scores reflect a decrease in salivary cortisol concentration after substance application. Separate *t*-tests compared change scores between odor groups for each of the two experimental saliva samples.

Central nervous face processing. For P100 and N170 ANOVAs with hemisphere, emotion and actor´s sex as within factors and odor as between subjects factor were calculated. For EPN ANOVA with hemisphere, site, emotion and actor´s sex as within and odor as between subjects factor was calculated. For LPP ANOVA with site, emotion and actor´s sex as within and odor as between subjects factor was calculated. For EPN and LPP emotion effects were followed by *Student´s t*-tests comparing affective with neutral faces. For P100, N170 and P300 emotion effects were followed by ANOVAs and pair-wise comparisons.

Central nervous scene processing. For P100, ANOVA with hemisphere, category and valence as within factors and odor as between subjects factor was calculated. For EPN, ANOVA with hemisphere, site, category and valence as within and odor as between subjects factor was calculated. For the components P300 and LPP, ANOVAs with site, category and valence as within and odor as between subjects factor were calculated. For EPN and LPP, valence effects were followed by *Student´s t*-tests comparing affective with neutral scenes. For P100 and P300, valence effects were followed by ANOVAs and pair-wise comparisons.

For reasons of simplicity only significant main effects, and interactions with the factors odor and emotion or valence are reported. Main effects are followed by pair-wise comparisons. Interactions with the factor odor were followed by *Student's t*-tests, interactions with the factor emotion or valence, respectively, were followed by ANOVAs. If necessary, Greenhouse-Geisser corrections of degrees of freedom were applied for violations of sphericity.

5.2.2 Results

5.2.2.1 Compound rating

Subjective ratings of the control and androstadienone solution revealed similar results. *t*-tests comparing the ratings of pleasantness (control: $M = 4.41$, $SD = 2.37$; androstadienone: $M = 4.05$, $SD = 1.96$), intensity (control: $M = 5.06$, $SD = 1.48$; androstadienone: $M = 4.86$, $SD = 2.27$) and familiarity (control: $M = 6.94$, $SD = 1.98$; androstadienone: $M =$

6.77, *SD* = 1.95) did not showed any significant differences in subjective perception of these olfactory qualities (all *ps* > .60).

5.2.2.2 Face ratings

ANOVA revealed no significant influence of androstadienone on subjective face ratings, all *ps* > .25. As expected, happy faces were rated as more positive, attractive and likable than neutral or angry faces. Angry faces were rated as more negative, less attractive and less likable than neutral faces. Furthermore, happy and angry faces were rated as more arousing than neutral faces and angry faces were rated as more arousing than happy faces. See Table 4 for details.

Valence. ANOVA revealed a significant main effect of emotion, $F(2, 98) = 320.89$, $p < .001$, $\eta_p^2 = .87$. Pair-wise comparisons revealed significant differences between emotional expressions, with happy faces rated as more positive than neutral faces and angry faces rated as more negative than neutral faces (all *ps* < .001)

Arousal. ANOVA revealed a significant main effect of emotion, $F(2, 98) = 27.17$, $p < .001$, $\eta_p^2 = .36$. Pair-wise comparisons revealed significant differences between emotional expressions, with happy and angry faces rated as more arousing than neutral faces (all *ps* < .001) and angry faces rated as more arousing than happy faces ($p = .050$).

Attractiveness. ANOVA revealed a significant emotion effect, $F(2, 98) = 88.85$, $p < .001$, $\eta_p^2 = .65$ and a significant Actor´s sex x Emotion interaction, $F(2, 98) = 1.42$, $p = .017$, $\eta_p^2 = .09$. Follow up *t*-test separately for each gender revealed for both, men and women, significant emotion effects, all $Fs(2,104) > 78.08$, all *ps* < . 001, all $\eta_p^2 > .59$. Pair-wise comparisons indicated that both, male and female happy faces were rated as more attractive and both male and female angry faces as less attractive than neutral faces (all *ps* < .001).

Sympathy. ANOVA revealed a significant main effect of emotion, $F(2, 98) = 155.31$, $p < .001$, $\eta_p^2 = .76$ and a significant Actor´s sex x Emotion interaction, $F(2, 98) = 4.53$, $p = .014$, $\eta_p^2 = .09$. Follow up t-test separately for each gender revealed for both, men and women, significant emotion effects, all $Fs(2,104) > 132.97$, all $ps < .001$, all $\eta_p^2 > .71$. Pair-wise comparisons indicated that both male and female happy faces were rated as more likable than neutral and both male and female angry faces were rated as less likable than neutral faces (all $ps < .001$).

Table 4. Mean face ratings of androstadienone and control groups, SD in brackets

	Androstadienone			Control		
	Angry	Happy	Neutral	Angry	Happy	Neutral
Arousal	4.66	5.07	3.66	4.97	5.25	3.93
	(1.44)	(1.48)	(1.27)	(1.29)	(0.93)	(1.15)
Valence	2.78	6.74	4.53	2.95	6.72	4.52
	(0.81)	(0.92)	(0.54)	(0.84)	(1.02)	(0.48)
Attractiveness	3,08	4.78	3.90	3.12	5.03	4.26
	(0.94)	(1.12)	(0.87)	(0.66)	(1.13)	(0.84)
Sympathy	2.98	6.23	4.33	2.94	6.33	4.49
	(1.07)	(1.25)	(1.02)	(0.81)	(1.51)	(1.10)

5.2.2.3 Scene ratings

As expected positive pictures were rated as more positive and negative pictures were rates as more negative than neutral pictures in all categories. In categories "social" and "non-social" positive and negative pictures elicited higher arousal ratings compared to neutral pictures. In the category "couples" positive pictures were rated as more arousing than negative or neutral pictures, which did not differ in arousal. See Table 5 and Table 6 for details.

Valence. A significant main effect of valence, $F(2, 98) = 507.69$, $p < .001$, $\eta_p^2 = .91$ and significant interactions of Category x Odor, $F(2, 98) =$

3.22, $p = .048$, $\eta_p^2 = .06$ and Category x Valence, $F(4, 196) = 34.83$, $p < .001$, $\eta_p^2 = .42$, were found. *t*-tests separately for each category revealed no significant influences of odors, all $ps > .06$. ANOVA testing valence effects reached significance for all categories; all $Fs(2, 100) > 311.53$, all $ps < .001$, all $\eta_p^2 > .85$. Pair-wise comparisons showed for all categories that positive scenes were rates as more positive than neutral scenes and negative scenes were rated as more negative than neutral scenes (all $ps < .002$).

Arousal. ANOVA revealed no significant odor effects, all $ps > .24$. ANOVA revealed a significant main effect of valence, $F(2, 98) = 47.19$ and a significant Valence x Category interaction, $F(4, 196) = 31.98$, $p < .001$, $\eta_p^2 = .40$. Follow up ANOVAs revealed for all categories a significant valence effect; all $Fs(2, 100) = 6.79$, all $ps < .001$, all $\eta_p^2 > .12$. Pair-wise comparisons showed for the social and non-social categories that negative and positive scenes were rated as significantly more arousing than neutral scenes, all $ps < .001$. Positive scenes in the category "couples" were rated as more arousing than negative or neutral scenes, all pair-wise comparisons $ps < .02$. Negative and neutral scenes did not differ in arousal ratings, $p = 1.0$.

Table 5. Mean scene ratings of the androstadienone group, SD in brackets.

	Social			Non-social			Couples		
	Negative	Positive	Neutral	Negative	Positive	Neutral	Negative	Positive	Neutral
Arousal	5.94	6.08	4.89	5.36	5.04	2.81	5.76	6.82	5.97
	(1.50)	(1.55)	(1.17)	(1.34)	(1.34)	(1.51)	(1.66)	(1.69)	(1.35)
Valence	2.23	7.61	5.76	2.16	7.13	4.41	2.18	7.45	6.82
	(1.00)	(1.07)	(0.90)	(0.85)	(0.99)	(1.16)	(0.86)	(0.93)	(0.93)

Table 6. Mean scene ratings of the control group, SD in brackets.

	Social			Non-social			Couples		
	Negative	Positive	Neutral	Negative	Positive	Neutral	Negative	Positive	Neutral
Arousal	6.14	6.35	5.03	5.76	5.34	3.29	6.10	6.29	6.21
	(1.25)	(1.21)	(1.14)	(1.52)	(1.29)	(1.65)	(1.37)	(1.41)	(1.04)
Valence	2.41	7.50	5.82	2.51	7.04	4.92	2.50	6.91	6.55
	(1.05)	(1.03)	(0.74)	(0.96)	(0.94)	(0.25)	(1.00)	(1.35)	(0.87)

5.2.2.4 Subjective mood ratings

Mood and anxiety. ANOVAs showed no odor effects on the mood variables of feeling focused, social, energetic, open, relaxed, sensual and irritated neither after the first nor after the second odor application, all $ps > .06$ (see 8.2.3 for details). ANOVA after the first odor application for feeling heavy revealed no significant odor effects (all $ps > .77$), but after the second odor application the interaction of Odor x Time reached significance, $F(3, 144) = 2.88$, $p = .047$, $\eta_p^2 = .06$, (see Figure 4). Following *t*-tests for each time point revealed no differences between groups (all $ps > .18$). For all mood variables, except for feeling relaxed ($p = .96$) and focused ($p = .09$), the main effect time reached significance after the first odor application, all $Fs(3, 144) > 3.50$, all $ps < .036$, all $\eta_p^2 > .06$. For all mood descriptors a significant main effect of time was found after the second odor application, all $Fs(3, 144) > 5.15$, all $ps < .004$, all $\eta_p^2 > .09$.

ANOVAs testing odor influences on state anxiety after the first and the second odor application revealed neither significant odor effects, all $ps > .43$, nor significant time effects, all $ps > .09$ (see 8.2.3).

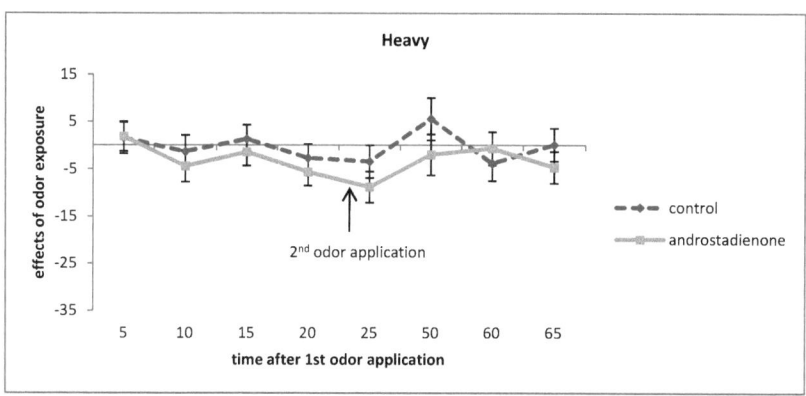

Figure 4. Mean odor effects on feeling heavy ± SEM in control (dark) and androstadienone (light) groups.

5.2.2.5 Alertness

Alertness. Student's *t*-tests of tonic and phasic alertness scores between odors demonstrated no influence of androstadienone on participants' task performance, tonic alertness: $p = .88$; phasic alertness: $p = .72$ (see Table 7).

Table 7. Mean reaction times of participant's median reaction times of tonic and phasic alertness values in androstadienone and control groups in ms, SD in brackets.

	Androstadienone	Control
Tonic alertness	244.9 (30.73)	246.20 (27.37)
Phasic alertness	- 8.21 (42.41)	-12.64 (47.52)

5.2.2.6 Physiological measurements

Heart rate. Figure 5 shows mean heart rate changes after odor application for the control and the androstadienone group. A significant main effect of time, $F(3, 132) = 7.68$, $p < .001$, $\eta_p^2 = .15$ and a significant

Time x Odor interaction, $F(3, 132) = 5.10$, $p = .005$, $\eta_p^2 = .10$, was found after the first compound application. Follow up t-tests for each time point between odor groups reached no significance all $ps > .16$. ANOVA testing the effect of the second odor application revealed a significant time effect, $F(3, 132) = 11.79$, $p < .001$, $\eta_p^2 = .22$, but no significant effects of odor (all $ps > .28$).

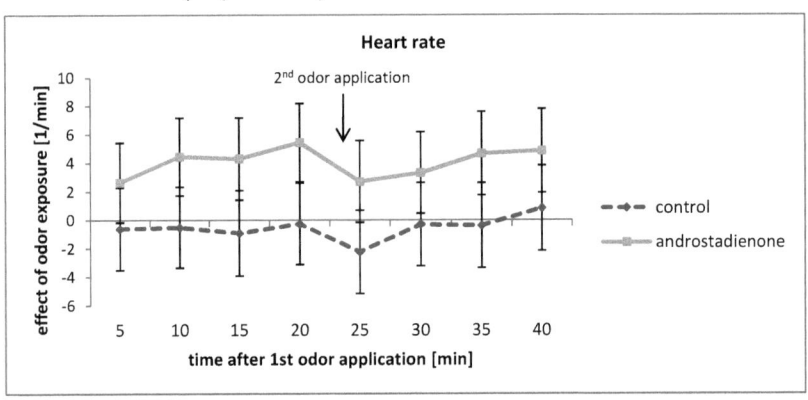

Figure 5. Heart rate changes ± SEM of odor exposure in the control and the androstadienone groups.

Skin conductance. Figure 6 shows mean changes of skin conductance after odor application for the control and the androstadienone group. A significant main effect of time, $F(3, 141) = 34.34$, $p < .001$, $\eta_p^2 = .42$ and a significant Time x Odor interaction, $F(3, 141) = 5.50$, $p = .009$, $\eta_p^2 = .42$, was found after the first odor application. Follow up t-tests for each time point between odor groups revealed no significant differences, all $ps > .07$. ANOVA testing the effect of the second odor application revealed a significant main effect of time, $F(3, 141) = 13.85$, $p < .001$, $\eta_p^2 = .23$, but no significant odor effects (all $ps > .37$).

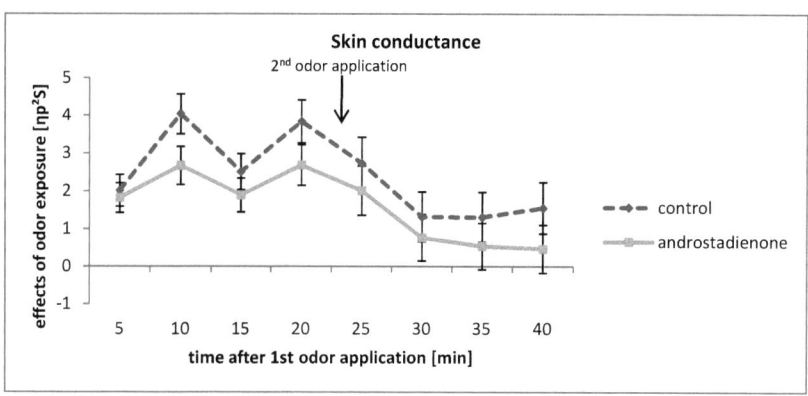

Figure 6. Skin conductance changes ± SEM of odor exposure in the control and the androstadienone group.

Respiration rate. Figure 7 shows mean changes of respiration rate after odor application for the control and the androstadienone groups. A significant main effect of time was revealed after the first, $F(3, 141) = 8.76$, $p = .001$, $\eta_p^2 = .16$ and the second odor application, $F(3, 141) = 4.16$, $p = .018$, $\eta_p^2 = .08$. No significant odor effects were found (all $ps > .29$).

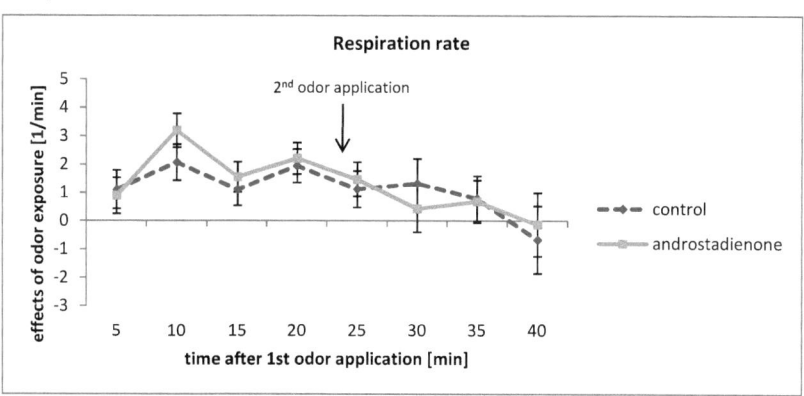

Figure 7. Respiration rate changes ± SEM of odor exposure in the control and the androstadienone group.

Cortisol. Figure 8 shows changes in cortisol concentrations after the first and the second odor application for the control and the androstadienone groups. *t*-test for group differences after the first odor application revealed a marginal significant difference between odor groups, $t(49) = 1.93$, $p = .059$. *t*-tests for the control group for differences from zero revealed significance for both samples, after the first application: $t(24) = 3.21$, $p = .004$; after the second application: $t(24) = 4.42$, $p < .001$, indicating a dropping cortisol level in the control group circa 20 and 60 minutes after baseline sampling. *t*-tests for the androstadienone group revealed no significant differences in cortisol concentrations from zero, all $ts < 1.58$, all $ps > .13$, indicating a rather stable cortisol levels after both androstadienone applications. No significant difference in cortisol levels between odor groups were revealed after the second odor application, $t(49) = .72$, $p = .71$.

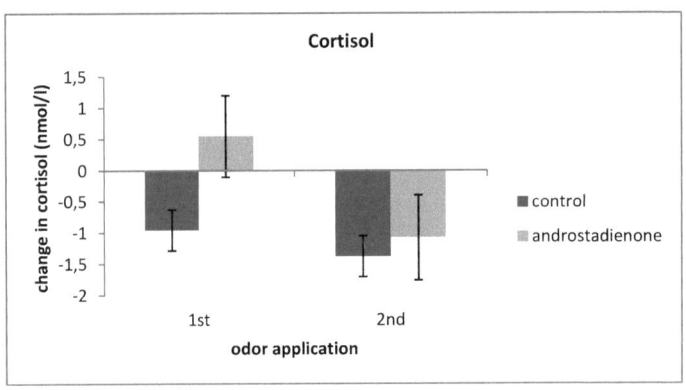

Figure 8. Change scores in salivary cortisol levels in nmol/l after the first and second exposure to control and androstadienone

5.2.2.7 Central nervous processing of faces

P100. ANOVA revealed no significant effects of odor, all $ps > .10$ (see Figure 9). A significant Emotion x Actor's sex interaction, $F(2, 98) = 4.24$, $p = .021$, $\eta_p^2 = .08$, was found. Following *t*-tests for each emotion demonstrated significant larger amplitudes to angry women ($M = 5.72$

µV, *SD* = 4.53) compared to angry men (*M* = 4.67 µV, *SD* = 4.94), *t*(50) = 2,25, *p* = .029, and a significant larger amplitudes to neutral men (M = 6.44 µV, SD = 4.71) compared to neutral women (*M* = 5.54 µV, *SD* = 4.34), *t*(50) = 2,05, *p* = .046. No significant actor´s sex differences were found for happy faces.

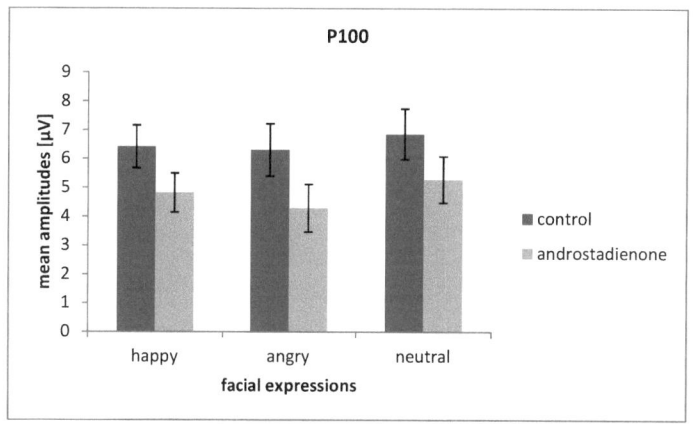

Figure 9. Mean P100 amplitudes ± SEM in response to happy, angry and neutral faces on electrode O2 in control (dark bar) and androstadienone groups (light bar).

N170. ANOVA revealed no significant influence of odor, all *p*s > .09 (see Figure 10). Also no main effects or interactions with the factor emotion revealed significance, all *p*s > .14.

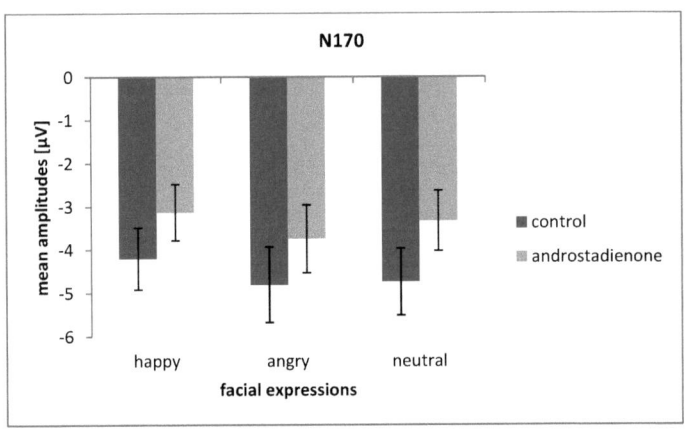

Figure 10. Mean N170 amplitudes ± SEM in response to happy, angry and neutral faces on the electrode P8 in control (dark bar) and androstadienone groups (light bar).

EPN. ANOVA revealed no significant effects of odor, all $ps > .08$ (see Figure 11), but showed a significant main effect of emotion, $F(2, 98) = 3.82$, $p = .028$, $\eta_p^2 = .07$ and a Emotion x Site interaction, $F(6, 294) = 5.52$, $p < .001$, $\eta_p^2 = .10$. Following *t*-tests revealed on PO7/PO8 significant EPNs to angry ($M = 5.51$ µV, $SD = 3.87$) and happy faces ($M = 5.38$ µV, $SD = 3.54$) compared to neutral faces ($M = 6.22$ µV, $SD = 3.89$), $t(50) = 2.27$, $p = .028$ and $t(50) = 2.68$, $p = .010$, respectively. On PO9/PO10 also significant EPNs to angry ($M = 2.24$ µV, $SD = 2.99$) and happy faces ($M = 2.62$ µV, $SD = 2.78$) compared to neutral faces ($M = 3.31$ µV, $SD = 3.06$), were revealed, $t(50) = 3.97$, $p < .001$ and $t(50) = 2.77$, $p = .008$, respectively. On O1/O2 only an EPN to happy ($M = 5.24$ µV, $SD = 3.90$) compared to neutral faces ($M = 6.14$ µV, $SD = 3.94$) was found, $t(50) = 2.64$, $p = .011$.

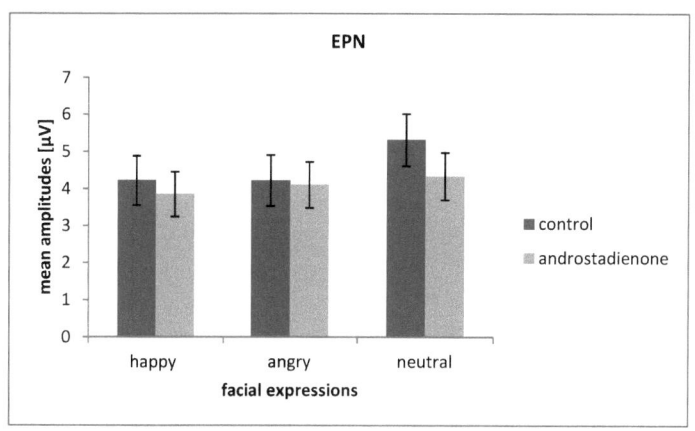

Figure 11. Mean amplitudes ± SEM in response happy, angry and neutral faces 280 – 310 ms post stimulus on the electrode PO8 in control (dark bars) and androstadienone groups (light bars).

P300. Figure 12 shows mean amplitudes on parietal sites for the control and the androstadienone groups. A significant Emotion x Odor effect, $F(2, 98) = 4.11$, $p = .022$, $\eta_p^2 = .08$, was demonstrated. Follow up *t*-tests for each emotion testing differences between odors revealed no significant effects, all ps >.34. In addition, significant interactions of Emotion x Site, $F(4, 196) = 3.17$, $p = .024$, $\eta_p^2 = .06$ and Emotion x Site x Actor´s sex, $F(4, 196) = 2.96$, $p = .027$, $\eta_p^2 = .06$, were found. Following ANOVAs for each site with emotion and actor´s sex as within factors revealed on Pz a significant main effect of emotion, $F(2, 100) = 3.84$, $p = .027$, $\eta_p^2 = .07$. Pair-wise comparisons showed significant larger amplitudes towards angry faces ($M = 4.33$ µV, $SD = 4.80$) compared to neutral faces ($M = 3.15$ µV, $SD = 4.84$), $p = .005$.

Additional explorative ANOVAs for the P300 component following the significant Emotion x Odor effect compared reactions to facial expressions within each group. An emotion effect was revealed for the control group, $F(2, 44) = 4.67$, $p = .017$, $\eta_p^2 = .18$. Pair-wise comparisons showed differences between reactions to happy and angry faces, $p = .003$, indicating a higher amplitude to angry compared to happy faces. No differences between emotional and neutral faces were revealed. For the androstadienone group no emotion effect was revealed, $p = .12$.

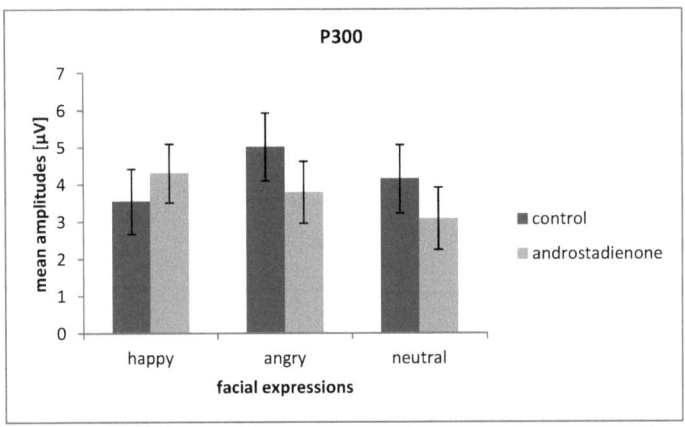

Figure 12. Mean amplitudes ± SEM on parietal sites (P3, Pz, P4) 320 – 360 ms post stimulus in response to happy, angry and neutral faces for the control and the androstadienone group.

LPP. ANOVA revealed no significant odor effects, all $ps > .08$ (see Figure 13). The main effect emotion reached significance, $F(2, 98) = 9.30$, $p < .001$, $\eta_p^2 = .16$. Pair-wise comparisons showed significant LPPs to angry ($M = 3.32$ µV, $SD = 4.35$) and happy faces (M = 2.47 µV, SD = 3.88) compared to neutral faces ($M = 1.66$ µV, $SD = 3.78$), $p < .001$ and $p = .027$, respectively.

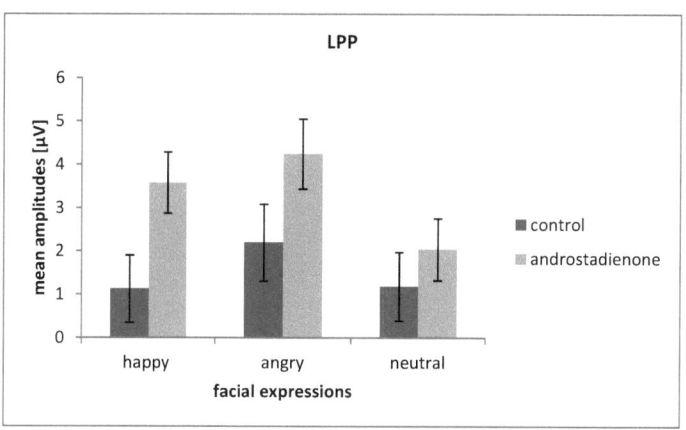

Figure 13. Mean amplitudes ± SEM in response happy, angry and neutral faces 400 -600 ms post stimulus on Pz in control (dark bars) and androstadienone groups (light bars).

5.2.2.8 Central nervous processing of scenes

P100. Analysis revealed no significant effects of odor, all $ps > .07$ (see Figure 14). Significant interactions of Category x Valence, $F(1, 196) = 11.03$, $p < .001$, $\eta_p^2 = .18$ and Hemisphere x Category x Valence, $F(4,196) = 6.22$, $p < .001$, $\eta_p^2 = .11$, were revealed. Separate ANOVAs revealed for both hemispheres a significant Category x Valence interaction, all $Fs(2, 400) > 4,78$, all $ps > .001$, all $\eta_p^2 > .08$. ANOVAs showed for both hemispheres significant valence effects for non-social scenes and in the left hemisphere a significant valence effect for couple scenes, all $Fs(2, 100) > 7.47$, all $ps < .001$, all $\eta_p^2 > .12$. Pair-wise comparisons revealed for both hemispheres larger amplitudes to non-social negative (left: $M = 6.95$ µV, $SD = 5.63$; right: $M = 6.22$ µV, $SD = 5.44$) compared to non-social positive (left: $M = 4.07$ µV, $SD = 5.56$; right: $M = 4.34$ µV, $SD = 5.48$) or neutral scenes (left: $M = 5.20$ µV, $SD = 4.65$; right: $M = 5.20$ µV, $SD = 5.63$), all $ps < .030$. In the left hemisphere also significant larger amplitudes were revealed for non-social neutral compared to non-social positive scenes, $p = .012$. Positive ($M = 5.67$ µV, $SD = 4.72$) and neutral scenes with couples ($M = 6.39$ µV, $SD = 5.44$) elicited only in the left hemisphere larger amplitudes compared to negative scenes with couples ($M = 3.93$, $SD = 4.42$), $p < .001$.

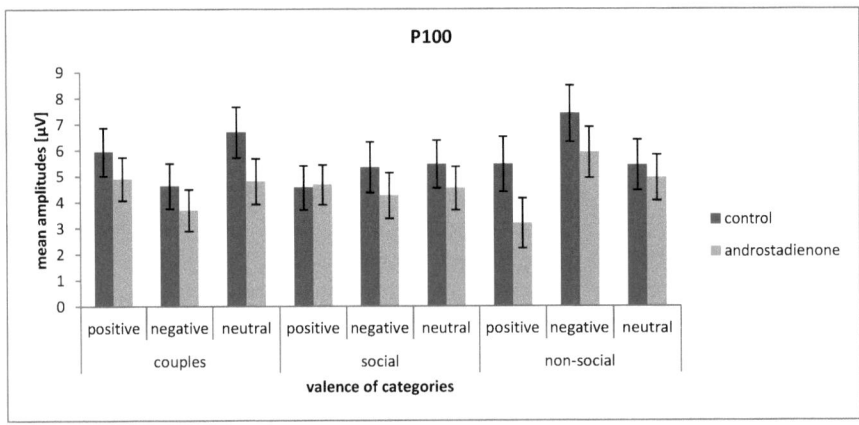

Figure 14. Mean P100 amplitudes ± SEM on O2 in response to positive, negative and neutral scenes of the categories couples, social and non-social in control (dark bars) and androstadienone groups (light bars).

EPN. ANOVA revealed a significant main effect of valence, $F(2,98) = 10.16$, $p < .001$, $\eta_p^2 = .17$ and a significant interaction of Hemisphere x Site x Odor, $F(3, 147) = 6.41$, $p = .002$, $\eta_p^2 = .12$ (see Figure 15). Following ANOVAs for each site with hemisphere as within and odor as between subjects factor demonstrated only for PO7/PO8 a significant Hemisphere x Odor interaction, $F(1, 49) = 6.40$, $p = .015$, $\eta_p^2 = .12$. Follow up *t*-tests between odors for each hemisphere revealed no significant effects, all $ps > .12$. Also, the Site x Category x Valence interaction reached significance, $F(12, 588) = 12.25$, $p < .001$, $\eta_p^2 = .20$. Following ANOVAs for each site with category and valence as within factors revealed for all sites significant interactions of category and valence, all $Fs(4, 200) = 9.64$, all $ps < .022$, all $\eta_p^2 > .06$ (see Figure 16).

Following *t*-tests on PO7/PO8 comparing affective with neutral scenes for each category revealed a significant EPN only to positive social (M = 6.18 µV, SD = 4.96) compared to neutral social scenes (M = 8.57 µV, SD = 4.52), $p < .001$. *t*-tests on O1/O2 also revealed only for positive social scenes (M = 4.41 µV, SD = 5.62) an EPN compared to neutral social scenes (M = 7.00 µV, SD = 4.83), $p < .001$. On PO9/PO10 for scenes with couples a significant EPN to positive (M = 2.60 µV, SD = 3.64) compared to neutral scenes (M = 4.51 µV, SD = 3.41) was found, $p < .001$. For social scenes a significant EPN to positive (M = 4.57 µV, SD = 4.40) compared to neutral scenes (M = 6.41 µV, SD = 3.59) was found. On O9/O10 for scenes with couples a significant EPN to positive (M = 3.26 µV, SD = 3.82) compared to neutral scenes (M = 4.35 µV, SD = 3.39) was found, $p = .006$. Also for social scenes a significant EPN to positive (M = 3.52 µV, SD = 4.17) compared to neutral scenes (M = 5.40 µV, SD = 3.54) was revealed, $p < .001$. In response to negative scenes in none of the categories significant EPNs compared to neutral scenes were demonstrated. Affective scenes in the category non-social elicited no EPN.

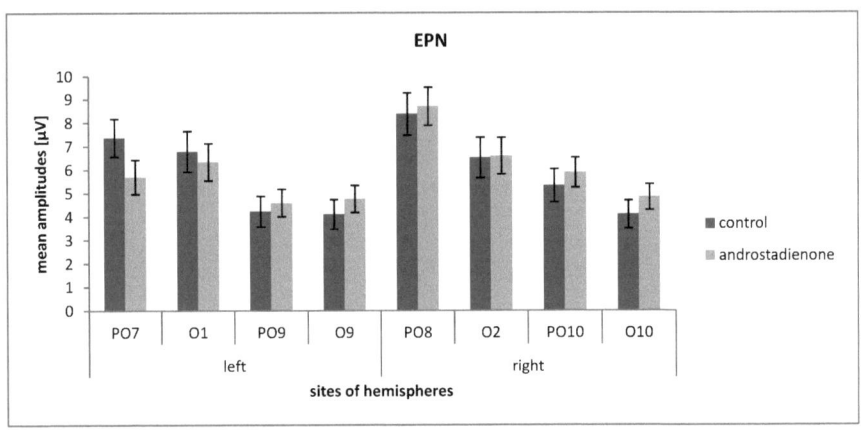

Figure 15. Mean amplitudes ± SEM 280 – 310 ms post stimulus in response to scenes in control (dark bars) and androstadienone groups (light bars).

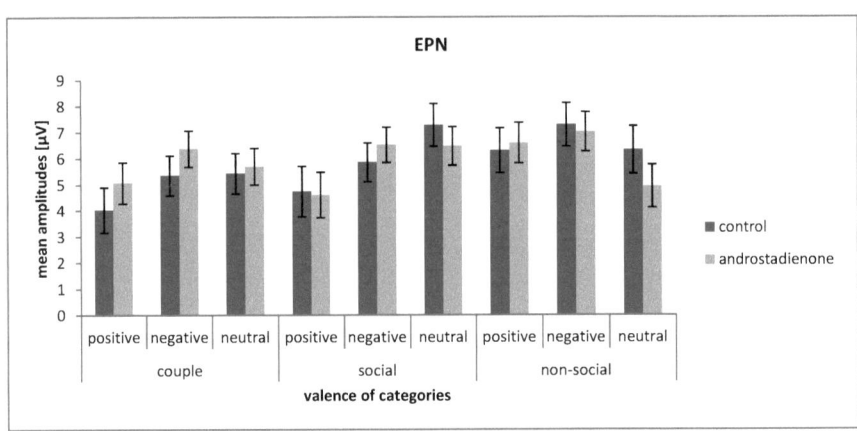

Figure 16. Mean amplitudes ± SEM 280 – 310 ms post stimulus on PO8 in response to positive, negative and neutral scenes of the categories couples, social and non-social in control (dark bars) and androstadienone groups (light bars).

P300. ANOVA revealed no significant odor effects, all *p*s > .14 (see Figure 17). A significant main effect of valence, $F(2, 98) = 4.40$, $p = .019$, $\eta_p^2 = .08$, a significant Category x Valence interaction, $F(4, 196) = 12.44$, $p < .001$, $\eta_p^2 = .20$ and a significant Site x Category x Valence interaction were found, $F(8,392) = 3.06$, $p = .032$, $\eta_p^2 = .06$. For each site ANOVAs with category and valence as within factors were calculated. For all sites significant Category x Valence interactions were found, all $Fs(4, 200) > 8.87$, all $ps < .001$, all $\eta_p^2 > .14$. ANOVAs for each category revealed on P3 a significant valence effect for the category "couples", $F(2, 100) = 25.20$, $p < .001$, $\eta_p^2 = .34$, with positive scenes ($M = 5.86$ µV, $SD = 4.34$) eliciting larger amplitudes than neutral ($M = 2.75$ µV, $SD = 3.76$) or negative scenes ($M = 2.62$, $SD = 3.69$), all $ps < .001$. Also on P3 a significant valence effect for the category "non-social", $F(2, 100) = 6.96$, $p = .002$, $\eta_p^2 = .12$, demonstrated larger amplitudes to negative ($M = 3.02$ µV, $SD = 5.04$) compared to positive ($M = .75$ µV, $SD = 5.33$) or neutral scenes ($M = 1.66$, $SD = 4.29$), all $ps < .001$. ANOVAs for each category revealed on Pz a significant valence effect for the category "couples", $F(2, 100) = 17.99$, $p < .001$, $\eta_p^2 = .27$, with positive scenes ($M = 6.64$ µV, $SD = 4.34$) eliciting larger amplitudes than neutral ($M = 2.80$ µV, $SD = 3.76$) or negative scenes ($M = 3.52$, $SD = 3.69$), all $ps < .001$. And a significant valence effect for the category "non-social", $F(2, 100) = 10.06$, $p < .001$, $\eta_p^2 = .17$, demonstrated larger amplitudes to negative (M = 2.90 µV, SD = 4.14) compared to positive (M = .47 µV, SD = 5.03) or neutral scenes ($M = 1.25$, $SD = 4.58$), all $ps < .001$. ANOVAs for each category revealed on Pz a significant valence effect for the category "couples", $F(2, 100) = 15.12$, $p < .001$, $\eta_p^2 = .23$, with positive scenes ($M = 7.52$ µV, $SD = 3.94$) eliciting larger amplitudes than neutral ($M = 5.11$ µV, $SD = 4.06$) or negative scenes ($M = 5.35$, $SD = 3.48$), all $ps < .001$. And a significant valence effect for the category "non-social", $F(2, 100) = 11.84$, $p < .001$, $\eta_p^2 = .19$, demonstrated larger amplitudes to negative ($M = 5.31$ µV, $SD = 4.74$) compared to positive ($M = 3.08$ µV, $SD = 5.22$) or neutral scenes ($M = 2.96$, $SD = 4.64$), all $ps < .001$.

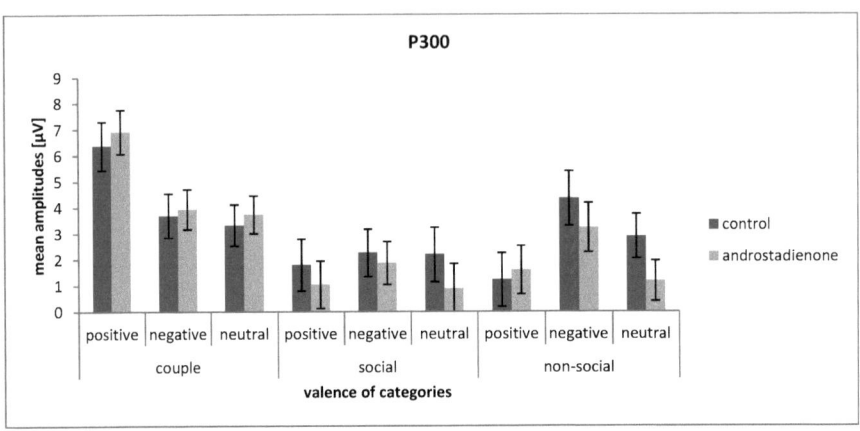

Figure 17. Mean P300 amplitudes ± SEM on Pz in response to positive, negative and neutral scenes of the categories couples, social and non-social in control (dark bars) and androstadienone groups (light bars).

LPP. ANOVA showed a significant main effect of valence, $F(2, 98) = 14.53$, $p < .001$, $\eta_p^2 = .23$, a Category x Valence interaction, $F(4, 196) = 23.56$, $p < .001$, $\eta_p^2 = .33$, a Site x Valence interaction, $F(4, 196) = 8.24$, $p < .001$, $\eta_p^2 = .14$ and a Site x Category x Valence interaction, $F(8, 392) = 4.51$, $p = .001$, $\eta_p^2 = .06$ (see Figure 18). ANOVAs for each site with category and valence as within factors revealed for all sites significant main effects of valence, all $Fs(2, 100) > 11.11$, all $ps < .001$, all $\eta_p^2 > .18$ and Category x Valence interactions, all $Fs(4, 200) > 16.80$, all $ps < .001$, all $\eta_p^2 > .25$. Following *t*-tests for each category testing differences between affective and neutral scenes showed an LPP to positive scenes with couples on all sites, all $Ts(50) > 6.15$, all $ps < .001$. An LPP to positive and negative social scenes were revealed on Fz and Cz, all $Ts(50) > 2.11$, all $ps < .040$. No LPPs were found for non-social affective compared to neutral scenes. A significant Category x Odor interaction, $F(2, 98) = 3.29$, $p = .042$, $\eta_p^2 = .06$, was followed by *t*-tests comparing odors within each category. Results did not reach significance, all $ps > .15$.

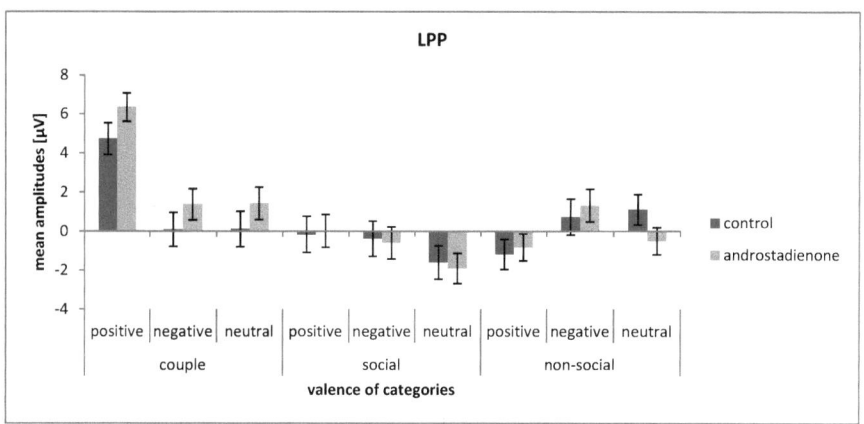

Figure 18. Mean amplitudes ± SEM 400 – 600 ms post stimulus on Pz in response to positive, negative and neutral scenes of the categories couples, social and non-social in control (dark bars) and androstadienone groups (light bars).

Explorative *t*-tests comparing categories within each odor group revealed for both groups an LPP to pictures with couples compared to social scenes: control group: $t(22) = 4.31$, $p < .001$; androstadienone group: $t(27) = 9.82$, $p < .001$; and non-social scenes: control group: $t(22) = 2.31$, $p = .030$; androstadienone group: $t(27) = 7.15$, $p < .001$ (see Figure 19).

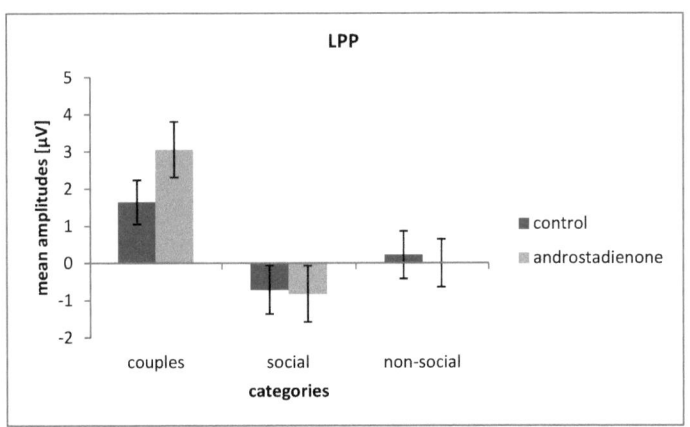

Figure 19. Mean amplitudes ± SEM 400 – 600 ms post stimulus on midline electrodes (Fz, Cz, Pz) in response to scenes with couples, social and non-social scenes, for the control (dark bars) and the androstadienone groups (light bars).

5.2.3 Discussion

Current results revealed no androstadienone effects on central nervous stimulus processing in women. Explorative analyses, however, suggested that androstadienone influences the processing of faces. In the control group a larger P300 amplitude to angry compared to happy faces indicated more attention allocation towards anger, while emotional and neutral faces were not processed differently. In contrast, the androstadienone group showed similar P300 amplitudes to happy and angry faces. This suggests that androstadienone eliminates emotional differentiation. This finding is in line with the reviewed literature, suggesting an androstadienone effect on higher cognitive processes, as the P300 a later portion of the visual ERP is reflecting exactly these processes. The P300 is known to enlarge with increased attention allocation towards a specific stimulus (Polich & Kok, 1995). Thus, the control group might have engaged more attention towards angry compared to happy faces, whereas androstadienone seemed to prevent this differential reaction. This might indicate that anger and happiness are equally important for a female observer, when she is inhaling the endogenous odorant. A conclusion, whether androstadienone reduced the attention allocated towards angry faces or enhanced attention towards happy faces is not possible, because differences between groups are not significant. Previous research suggested that androstadienone is able to enhance attention (Hummer & McClintock, 2009; Lundström et al., 2003), rather suggesting an enhancement of attention allocation towards happy faces than a decrease in attention towards anger. However, other attention measurements in the current study did not support this notion. Ratings of the feeling of being focused, an explicit measure of attention, were not modulated by androstadienone exposure. Likewise the alertness task performance was also not affected by androstadienone exposure. These results suggest no attentional modulation by androstadienone on a behavioral level. Still, this does not contradict the notion that androstadienone might modulate attention on a central nervous level, especially because emotional stimuli were not involved in behavioral attention measures. Therefore, as previously concluded by Hummer and McClintock (2009), androstadienone might rather affect the processing of affective than emotionally neutral information.

Usually, the anger expression is a significant stimulus as former studies have proven by behavioral and psychophysiological measurements (Öhman et al., 2001; Schupp et al., 2004). Also in the current study and in line with Kolassa et al. (2007), angry compared to neutral faces elicited an enhanced P300 amplitude independently of the applied odor. Other studies have shown that indeed meaningful faces elicit increased P300 amplitudes compared to less important faces

(Meijer, Smulders, Merckelbach, & Wolf, 2007; Paller et al., 2003). Hence, for our control group anger might have been more important than happiness, whereas androstadienone might have interfered with the outstanding role of anger and eliminated its preferred processing in the brain. But if androstadienone indeed eliminated the anger superiority, why this is not reflected by the EPN and LPP components? EPN and LPP specifically reflect the differences between processing of affective compared to neutral stimuli but not between affective stimuli. As mentioned above, rather than modulating the relation between emotional and neutral stimuli the endogenous odor may affect especially the processing of positive and negative emotional information.

An alternative mechanism for androstadienone effects may be its influence on subjective arousal. The P300 amplitude is also known to increase with increased arousal elicited by the stimulus material (Polich & Kok, 1995; Rozenkrants, Olofsson, & Polich, 2008). In the current study, angry faces were rated as more arousing than happy faces, which could explain the enhanced amplitude to angry compared to happy faces in the control group. However, both angry and happy faces were rated as more arousing than neutral ones, which was not reflected by P300 amplitudes. Also the androstadienone group rated angry faces as more arousing than happy faces, which was not reflected by the P300 amplitude. Moreover, physiological measurements did not reveal an androstadienone influence on physiological arousal. Therefore, an androstadienone related change in the arousal level seems to be an unlikely mechanism for the reported P300 changes.

Explorative findings have to be interpreted with care, first, because planned statistical analyses did not reveal any influence of androstadienone on central nervous attentional processes, and second, a subsequent study in our laboratory (unpublished data) could not replicate androstadienone effects on the P300 component. In this unpublished experiment we tested female's cortical reactions to similar emotional faces with angry, happy, neutral and also sad and fearful faces. However, no emotional differentiation in P300 amplitudes in the control group, and no androstadienone related modulation of the P300 component were revealed. Moreover, the emotional difference between angry and happy faces in P300 amplitudes found in the current study has never been reported in previous studies. Therefore, this seems to be a rather labile effect.

Still, the current paradigm and chosen stimuli elicited expected results showing no methodological problems in EEG assessment. In accordance with other studies, faces elicited emotionally unaffected P100 and N170 amplitudes (Eimer & Holmes, 2002; Eimer, Holmes, & McGlone, 2003; Holmes, Vuilleumier, & Eimer, 2003). Both happy and angry faces compared to neutral faces elicited pronounced EPNs and LPPs (Sato et al., 2001; Schupp, Junghöfer, Weike, & Hamm, 2003b; Schupp et al., 2004). In accordance, emotional faces were rated as more arousing than neutral faces. Moreover, compared to neutral pictures, positive scenes elicited an EPN and both, positive and negative scenes an LPP. Similar patterns of visual emotional information processing has been found repeatedly (Cuthbert et al., 2000; Mühlberger et al., 2009; Sato et al., 2001; Schupp et al., 2003a; Schupp et al., 2004). In line, positive compared to neutral pictures in all categories elicited more subjective arousal. One unexpected result was that negative compared to neutral pictures did not elicit an EPN. Valence ratings showed a more negative rating of negative compared to neutral scenes in all categories, and arousal ratings of negative scenes in the categories "social" and "non-social" showed higher scores compared to neutral scenes. However, similar arousal ratings of neutral and negative pictures in the category "couples" were revealed. As stimulus arousal level contributes to EPN this may cause the lacking EPN to negative compared to neutral scenes.

Furthermore, control measurements suggest correct odor application and appropriate experimental procedure. This is supported by maintained levels of salivary cortisol in the androstadienone compared to the control group, which is consistent with former findings (Wyart et al., 2007). Participants rated control and androstadienone solution as iso-intense as well as similar pleasant and familiar, suggesting that women could not discriminate between control and test solutions and were not able to consciously detect the endogenous odor.

However, the current study could not replicate reported effects of androstadienone on heart rate, skin conductance and respiration rate. But previous studies also reported no effects on physiology, which likely originate from a combination of methodological, stimulus application, task and analysis differences. An increase in physiological arousal related to androstadienone exposure was found by Bensafi and colleagues (2003, 2004), Lundström and Olsson (2005) and Wyart et al. (2007), but opposite androstadienone effects, like decreases in respiratory and heart rate, lowering of skin conductance and an increase of body temperature, were reported by Grosser and colleagues (2000). The most obvious difference between the mentioned experiments, which might account for this inconsistency, is again of contextual nature. In all studies reporting physiological enhancing effects exclusively a male experimenter presented androstadienone to female participants. Although, Grosser et al. (2000) did not explicitly mention the sex of the experimenter, this seems likely to account for the inconsistent findings. However, if Grosser et al. (2000) indeed tested women by a female experimenter, as I did in this thesis, they still found androstadienone related modulations in physiology in contrast to our studies. Their experimental setting was similar to ours: besides measuring physiology, EEG and also several questionnaires were assessed. However, odors were delivered directly to the VNO, whereas in our studies odor solutions were applied onto the upper lip of participants. Another reason might be the hormonal status of participating women. Our female participants took hormonal contraceptives, whereas women tested in all other physiological studies did not. Also the application method differed between studies. Bensafi et al. (2003), reporting increased skin conductance and respiration rate, applied androstadienone in crystal form. Other studies assessed physiology with additional physiological measures, like finger pulse, ear pulse, skin temperature, blood pressure, body movements, abdominal respiration and thoracic respiration (Bensafi, Brown et al., 2004; Bensafi et al., 2003; Wyart et al., 2007), then calculated *Z*-scores to reduce between subjects variability and pooled these measurements into an arousal index. To exclude statistical differences to be responsible for negative results in the current study, we also expressed physiological data as *Z*-scores and calculated ANOVAs with each physiological measure and time as within and odor as between subjects factors. However, odor effects did not reach significance (all *p*s > .66). Then, before merging all collected physiological measurements (skin conductance, thoracic respiration and heart rate) into an arousal index, we calculated *Person* tests to assure correlations between the physiological variables. However, required significant positive correlations were not revealed and prevented the calculation of the composite physiological arousal index.

Moreover, in our laboratory we could not replicate previously detected mood effects of androstadienone. Although this is in line with Hummer and McClintock (2009), it contradicts others (Bensafi et al., 2003; Grosser et al., 2000; Jacob & McClintock, 2000; Lundström et al., 2003). Interestingly, theses effects were found with a male experimenter, the opposite sex, but also with a female experimenter (Villemure & Bushnell, 2007). Therefore, the experimenter´s sex explanation seems unlikely to account for negative findings on mood in the current thesis. Still, a man conducting experiments with women seems to be an essential contextual cue for androstadienone effects on physiology. Another important factor in inducing androstadienone effects on mood and physiology might be the hormonal status of female participants. In contrast to former studies, female participants in the current study used hormonal contraceptives.

Moreover, no significant change in alertness tasks performance was found, which is in line with results reported by Lundström and colleagues (2005). Their task consisted of adjusting a smaller square inside a constantly moving larger square at all times for 20 minutes. In our task women were asked to react as fast as possible to an appearing square (with or without warning cue) by pressing one button. Both tasks aimed to measure the ability to maintain attention over a longer time period without demanding higher cognitive performance. This suggests that the lack of significant effects in the sustained attention tasks is rather related to the task at hand, than to the general inability of androstadienone to affect attention related behavioral responses.

One support for this notion comes from Lundström et al. (2003), who reported an androstadienone related enhanced subjective feeling of being focused. However, current experiment did not replicate this finding. It has to be mentioned that, in contrast to Lundström et al. (2003), the probably significant contextual cue, a male experimenter, was not present. Moreover, female participants took hormonal contraceptives, whereas Lundström's participants did not. Although Lundström and colleagues (2003) replicated this attention effect across three independent experiments it has not been replicated by another research group. Moreover, like this thesis, Jacob and McClintock (2000) and Jacob et al. (2002) failed to show an effect on subjective alertness in women inhaling androstadienone, although a male experimenter was present during testing. Notably, in these studies subjective attention was measured by a different questionnaire, than by Lundström (2003). Nevertheless, androstadienone related effects on subjective feeling of being attentive remain questionable until these effects are independently replicated.

Clearly one limitation of the current study is that only women using hormonal contraceptives were included, whereas most other studies reporting androstadienone effects tested only women, who were not on hormonal birth control. We opted for this inclusion criterion because olfactory sensitivity to androstadienone seemed not to be susceptible by the use of oral contraceptives. Lundström et al. (2003) investigated both, women using and not using hormonal contraceptives and tested the absolute detection threshold for androstadienone. Authors did not yield evidence that the use of oral contraceptives affects olfactory sensitivity for androstadienone. Moreover, behavioral measures were reported to be influenced by human sweat especially in women taking hormonal contraceptives. Thorne and colleagues (2002) reported improved attractiveness ratings of male faces by females exposed to male axillary secretions. These subjective evaluations of other persons were even better in women taking hormonal contraceptives. However, the fact that the current study did not replicate androstadienone effects on mood and some physiological measures indicates that androstadienone related effects at least on these reactions may indeed depend on women´s reproductive state.

Maybe the mask odor clove might have interfered with contraceptives and androstadienone effects. One study testing female´s androstadienone sensitivity, compared their performance between menstrual phases (Lundström, McClintock et al., 2006). Women in their fertile phase, i.e. day 7–15 post menses onset, were significantly more sensitive to androstadienone than women in their unfertile phase and women using oral contraceptives. Moreover, contraceptive users were more sensitive to the environmental odor, than to the social odor androstadienone. As we used the environmental odor clove to mask the endogenous odor in our studies, the heightened sensitivity to clove oil in the oral contraceptive users might have interfered with androstadienone effectiveness. This, among other reasons, may explain the lacking significance of androstadienone effects on some physiological, mood or central nervous measurements.

Moreover, testing androstadienone effects in a between-subjects design might be problematic regarding reactions towards olfactory stimuli. Indeed, groups were matched for age, but individual variability in sensitivity to androstadienone was not controlled for. Olfactory sensitivity can be influenced by sexual orientation, sensitization, experience, genetic determination and menstrual synchrony (Boyle et al., 2006;

Bremner et al., 2003; Dorries et al., 1989; Keller et al., 2007; Knaapila et al., 2008; Lubke et al., 2009; Morofushi et al., 2000; Wysocki & Beauchamp, 1984; Wysocki et al., 1987). All these factors may uncontrollably manipulate effects of the endogenous odor.

In general, also the overwhelming experimental setting in the current study may be responsible for missing androstadienone related modulations. In the current experiment, participants had to sit in a small chamber and were confronted with many different tasks over a time span of about two hours. This input overload and the long time period might have caused irritation, distraction and fatigue. This in turn, might have interfered with small and labile androstadienone effects.

Taken together, this study corroborates earlier findings of androstadienone effects on women's cortisol level, however, failed to replicate effects on heart rate, skin conductance, respiration rate, mood, alertness and subjective attention. Moreover, planned statistics did not reveal androstadienone related influences on central nervous stimulus processing. However, explorative analyses suggested that androstadienone influences the processing of faces on a later portion of central nervous processing reflecting cognitive processes. Androstadienone might therefore be able to change attention allocation on a subconscious level. Still, effects have to be further investigated with women not taking hormonal contraceptives using a within-subjects design and fewer distracting variables.

5.3 Study III: cortical reactions in men

Influences of androstadienone on attention related behavioral reactions in men and women were suggested in the first experiment of this thesis. Especially men reacted to androstadienone exposure with a modulated approach tendency towards emotional faces. Explorative analyses in study II suggested androstadienone effects on central nervous face processing in women. In the current study it was sought for mechanisms underlying androstadienone effects in men. Especially it was sought for androstadienone related changes in attentional processes, which might have led to previous findings. By electroencephalography brain reactions to emotional face were examined. Reviewed literature suggests an androstadienone related influence on higher cognitive processes. Therefore, we hypothesized changes in ERP components at a later stage of stimulus processing, like the P300, during androstadienone exposure. As previous literature reported that body odors activate the brain´s fear network and study I revealed an enhanced attention allocation towards anger, larger P300 amplitudes in response to angry facial expressions were expected.

5.3.1 Methods

5.3.1.1 Subjects

Subjects were students and employees recruited through flyers on the campus of the University of Pennsylvania. Twenty heterosexual men, non-smokers and right handed aged between 18 and 32 years ($M = 25.1$, $SD = 3.6$) participated in this study. Subjects with nasal congestion or psychological and physiological diseases were excluded. All had a self-reported normal sense of smell, verified by an olfactory discrimination test (MONEX-40) (Albrecht et al., submitted). All participants defined their sexual orientation as exclusively heterosexual according to the Kinsey scale (Kinsey et al., 1953). Eleven men described themselves as Caucasian or White, three men described themselves as African-American or Black and one man described himself as Asian. Four men were excluded because of dizziness during EEG recording or recording problems, resulting in 16 subjects to include in further analyses. Subjects provided written informed consent (see 8.3.1), as required from the Institutional Review Board at the *Monell Chemical Senses Center* and got paid 50 USD for compensation.

5.3.1.2 Compounds

The androstadienone solution consisted of a 250 µM concentration of androstadienone (Steraloid Inc., London; purity > 98%) in propylene glycol (Sigma Aldrich; purity > 99%) with an odor mask consisting of 1 % eugenol (Sigma Aldrich, purity > 99%). The control substance consisted of propylene glycol with 1% eugenol. Same concentrations were used in our previous experiments and in earlier studies (Jacob & McClintock, 2000; Lundström et al., 2003; Olsson et al., 2006). Both solutions were applied to the subject's nose via a constant air flow (2 l/m) using a custom built olfactometer (Boesveldt et al., 2010; Lundström, Gordon, Albrecht, Alden, & Boesveldt, 2010).

5.3.1.3 Stimulus presentation

Three cartoon faces with a happy, angry and neutral expression were used in this study (see Figure 20). Stimuli were displayed with the software *EPrime* (*Psychology Software Tools*, Inc., Pittsburgh, PA, US) on a 19-inch computer screen (resolution: 1024 x 768 pixels) about one meter in front of the participants with a size of 739 x 739 pixels against a black background. The neutral face was presented 80 times; the emotional faces were presented 40 times each, resulting in 160 pictures in total. Stimuli were presented in a randomized order for one second followed by an emotional identification task. Participants had to identify the emotion, which was expressed on the previous face by a key press with their index finger of their dominant hand. Following subject´s response a blank screen appeared for 1.4 – 2.0 seconds. The total picture presentation took about 10 minutes.

Figure 20. Presented cartoon faces with angry, happy and neutral expression

5.3.1.4 Measurement of event related brain potentials (ERPs)

The EEG was continuously recorded by the software *ActiView* (*BioSemi*, Amsterdam, Netherlands) with a sampling rate of 512 Hz from 32 FLAT active Ag-AgCl-electrodes (*BioSemi* , Amsterdam, Netherlands) mounted on a flexible cab according to the international 10/20 system (Fp1, AF3, F7, F3, FC1, FC5, T7, C3, CP1, CP5, P7, P3, Pz, PO3, O1, Oz, O2, PO4, P4, P8, CP6, CP2, C4, T8, FC6, FC2, F4, F8, AF4, Fp2, Pz, Cz) and the left and right mastoids (M1, M2). Vertical (above and below both eyes) and horizontal electrodes (at outer canthi of both eyes) recorded eye movements. The recorded datasets were processed off-line with the software *Brain Vision Analyzer Version 2.0* (*Brain Products*, Munich, Germany). First, EEG data were low-pass and high-pass filtered (0.1 Hz and 30 Hz, respectively) and additionally a 60 Hz notch filter was used. Afterwards, data were re-referenced to linked mastoids (M1, M2). Then, epochs from 100 ms before until 700 ms after picture onset were extracted and corrected for blink artifacts using a digital ocular correction

(Gratton et al., 1983). Baseline correction was performed using the 100 ms pre-stimulus. The automatic search for artifacts used maximum allowed amplitudes of ± 50 µV and a maximum voltage step between data points of 50 µV. Artifacts were removed for each individual channel. Then epochs were averaged separately for each facial expression, actor's sex, channel and participant. P100 was searched as a local positive maximum between 50 ms and 130 ms post stimulus on O1 and O2 electrodes. N170 was searched as a local negative maximum between 100 ms and 200 ms post stimulus on P7 and P8 electrodes. The EPN was calculated after visual inspection of grand averages over all participants and conditions as mean activity between 220ms and 260ms post stimulus over left hemispherical (PO3, P7 and O1) and analogous right hemispherical electrodes (PO4, P8 and O2). The LPP was estimated after visual inspection of grand over all averages as mean activity between 600 ms and 700 ms post stimulus over midline electrodes (Fz, Cz, Pz). The P300 was calculated on P3, Pz and P4 electrodes as mean amplitude between 320 and 360 ms post stimulus (c.f. Kolassa, Kolassa, Musial, & Miltner, 2007).

5.3.1.5 Procedure

A within-subjects design was used in such that each subject underwent two sessions, one control and one androstadienone session, on two separate but not consecutive days. Sequence of odors was randomized between subjects. A female experimenter completed all interactions with the participants. After filling out the written informed consent (see 8.3.1) EEG cap and electrodes were mounted. Then, the cannula for odor application was attached and subjects conducted ten training trials to familiarize with the picture presentation and the following emotion identification task. Next, air flow was started and participants had to rate the odor, either androstadienone (in the androstadienone session) or the control substance (in the control session), for intensity, pleasantness and familiarity on 100 millimeter visual analog scales (see 8.4). Afterwards, the participants were asked to follow the instructions presented on the computer screen, while cortical reactions were continuously recorded. After the EEG recording was finished, subjects had to rate all presented faces for intensity and pleasantness on 100 millimeter visual analog scales (see 8.4), while smelling the applied odor. Finally, the experimenter disconnected all devices and conducted the odor identification test to control subject´s sense of smell. Subjects got paid after the final session.

5.3.1.6 Statistical analysis

Ratings. For subjective pleasantness and intensity ratings of emotional faces ANOVAs with repeated measures with odor and emotion as within subjects factors were calculated. For odor ratings of pleasantness, intensity and familiarity *Student´s t*-tests were calculated between odors.

Central nervous face processing. ANOVAs with repeated measures of averages of participants´ brain reactions were calculated. The effects of hemisphere, emotion and odor as within subjects factors were tested on P100 and N170. For EPN the factor site was additionally included. For P300 and LPP ANOVAs with site, emotion and odor as within subjects factors were calculated. Significant main effects were followed by pair-wise comparisons, significant interactions by *Student´s t*-tests. For reasons of simplicity only main effects and interactions including the factors odor and/or emotion are reported. Data analyses were performed with SPSS version 11.0. Alpha level was set at $p = .05$ to define a significant result. If necessary, Greenhouse-Geisser corrections of degrees of freedom were applied for violations of sphericity.

5.3.2 Results

5.3.2.1 Ratings

Odor ratings. See Table 8 for subjective odor ratings of intensity, pleasantness and familiarity. *t*-test revealed no significant differences between subjective ratings of androstadienone and control odor in intensity, pleasantness and familiarity, all $ps > .30$. This indicates that participants were consciously not able to discriminate between the two applied solutions.

Table 8. **Intensity, pleasantness and familiarity ratings of androstadienone and the control solutions (SD in brackets).**

	Androstadienone	Control
Pleasantness	64.9 (12.1)	67.1 (12.4)
Intensity	26.8 (11.7)	30.4 (15.5)
Familiarity	50.3 (21.3)	45.6 (26.0)

Facial rating. Table 9 shows pleasantness and intensity ratings of neutral, angry and happy cartoon faces. ANOVA for pleasantness ratings did not reveal a significant main effect of odor, $p = .70$, nor a significant Odor x Emotion interaction, $F(2, 30) = 1.63$, $p = .217.001$, $\eta_p^2 = .10$, indicating no androstadienone influence on subjective pleasantness ratings of faces. A significant main effect of emotion was demonstrated, $F(2, 30) = 207.70$, $p < .001$, $\eta_p^2 = .93$. Pair-wise comparisons showed significant differences between emotional and neutral faces, all $ps < .001$. The happy face was rated as more pleasant than angry and neutral faces, whereas the angry face was rated as less pleasant than happy and neutral faces.

ANOVA for intensity ratings did not reveal a significant main effect of odor, $p = .11$, nor a significant interaction of Odor x Emotion, $F(2, 30) = .49$, $p = .578$, $\eta_p^2 = .03$, indicating no androstadienone influence on subjective intensity ratings of faces. As expected, a significant main effect of emotion was found, $F(2, 30) = 58.79$, $p < .001$, $\eta_p^2 = .80$. Pairwise comparisons revealed significant differences between the intensity of emotional and neutral faces. Happy and angry faces were rated as more intense than the neutral face, all $ps < .001$. Happy and angry faces did not differ significantly in intensity ratings, $p = .07$.

Table 9. Pleasantness and intensity ratings of angry, happy and neutral cartoon faces, while smelling androstadienone or control solution (SD in brackets)

	Androstadienone			Control		
	Angry	Neutral	Happy	Angry	Neutral	Happy
Pleasantness	15.4 (8.9)	49.9 (7.5)	76.5 (9.1)	13.3 (7.1)	52.1 (8.5)	77.4 (10.0)
Intensity	54.6 (20.1)	14.1 (8.0)	51.6 (17.9)	59.9 (20.7)	15.6 (7.9)	53.8 (18.5)

5.3.2.2 Event related brain potentials

P100. Analyses testing the odor effect on the first positive peak revealed no significant effects including the factors odor or emotion, all $ps > .07$ (see Figure 21).

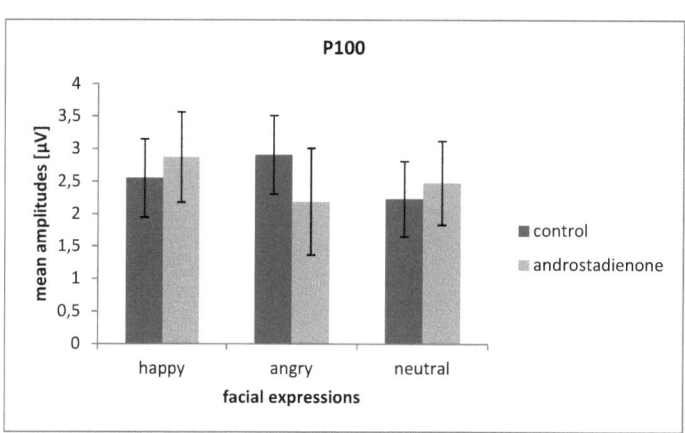

Figure 21. Mean P100 peaks ± SEM in response to happy, angry and neutral faces on O2 in control (dark bar) and androstadienone groups (light bar).

N170. Analyses testing the face specific negative peak revealed no significant effects of odor, all ps > .13 (see Figure 22). A significant Hemisphere x Emotion interaction, $F(2, 30) = 6.93$, $p = .005$, $\eta_p^2 = .32$, was followed by *t*-tests for each hemisphere. Only for the right hemisphere a significant difference between angry ($M = -4.11$ µV, $SD = 3.64$) and neutral ($M = -2.91$ µM, $SD = 2.57$), $p = .019$ faces was revealed.

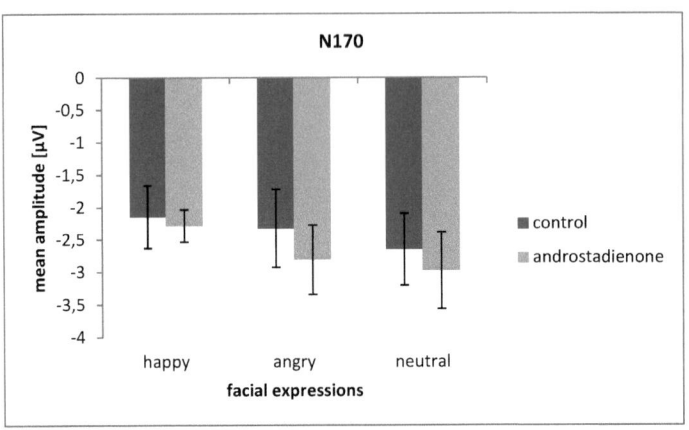

Figure 22. Mean N170 peaks ± SEM in response to happy, angry and neutral faces on P8 in control (dark bar) and androstadienone groups (light bar).

EPN. ANOVA revealed a significant Odor x Site x Emotion interaction, $F(2, 30) = 4.31$, $p = .009$, $\eta_p^2 = .22$. ANOVAs for each site with odor and emotion as within factors revealed no significant effects, all ps > .05 (see Figure 23). Explorative ANOVAs for each odor with site and emotion as within factors showed only for the control condition an interaction between Site and Emotion, $F(4, 60) = 3.40$, $p = .023$, $\eta_p^2 = .19$. *t*-tests for each site to test differences between emotional and neutral faces revealed no effects, all ps > .29.

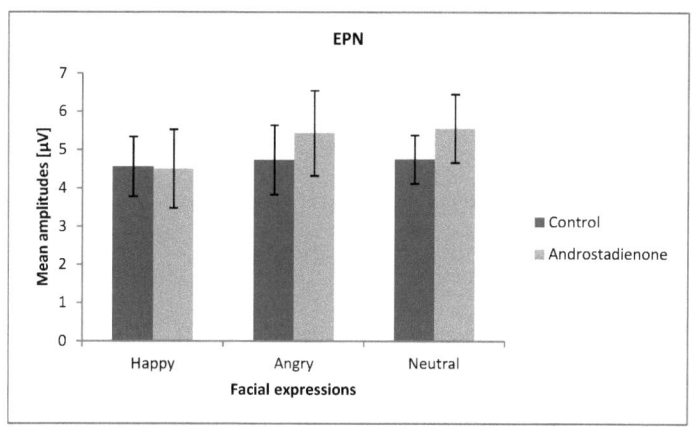

Figure 23. Mean EPN amplitudes ± SEM 220 – 260 ms post stimulus on P8 in response to happy, angry, and neutral faces in control (dark bar) and androstadienone groups (light bar).

P300. ANOVA revealed a significant main effect of odor, $F(1, 15) = 7.88$, $p = .013$, $\eta_p^2 = .34$, indicating a higher amplitude in the androstadienone compared to the control odor session (see Figure 24). No interactions with the factor odor reached significance, all ps > 12.

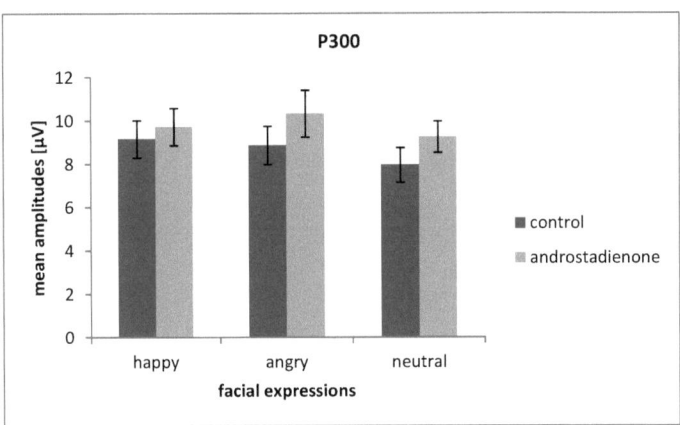

Figure 24. Mean P300 amplitudes ± SEM 320 - 360 ms post stimulus on Pz in response to happy, angry and neutral faces in control (dark bar) and androstadienone groups (light bar).

LPP. No significant effects of odor were revealed, all *p*s > .23 (see Figure 25). ANOVA revealed a significant main emotion, $F(2, 30) = 4.03$, $p < .001$, $\eta_p^2 = .21$ and a significant Site x Emotion interaction, $F(4, 60) = 5.17$, $p = .009$, $\eta_p^2 = .26$. Following *t*-tests for each site calculating differences between emotional and neutral faces revealed only for Pz a significant LPP in response to happy ($M = 5.36$ µV, $SD = 2.24$) compared to neutral faces ($M = 3.37$ µV, $SD = 2.43$), $t(15) = 3.23$, $p = .006$. No differences were reveled between LPP amplitudes in response to angry ($M = 3.93$, $SD = 2.12$) and neutral faces, $p = .32$.

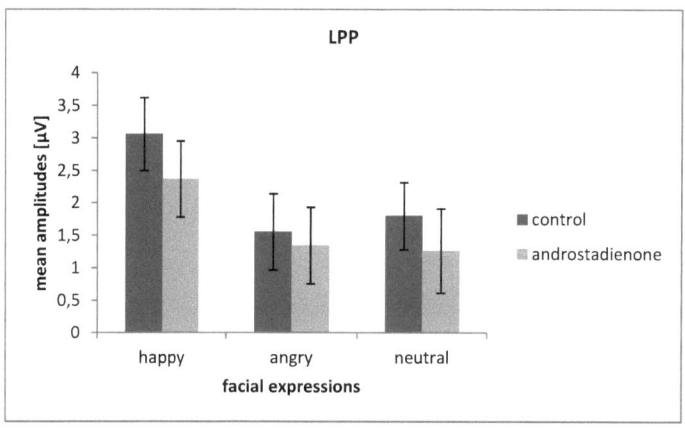

Figure 25. Mean LPP amplitudes ± SEM 600 - 700 ms post stimulus on Pz in response to happy, angry and neutral faces in control (dark bar) and androstadienone groups (light bar).

5.3.3 Discussion

Androstadienone modulated central nervous face processing in men. With androstadienone faces elicited larger P300 amplitudes compared to the control condition. The P300 as a later component of the ERP curve reflects attentional processes, which is in line with our expectations and several former studies suggesting that androstadienone might modulate higher cognitive processes (e.g.

Hummer & McClintock, 2009; Lundström, 2005; Lundström et al., 2003; e.g. Lundström & Olsson, 2005). Moreover, former studies showed that meaningful faces elicit a stronger P300 than unimportant faces (Meijer et al., 2007; Paller et al., 2003), which suggests that presented faces become more vital with smelling androstadienone. This might be explained by the fact that the sweat compound signals the olfactory presence of another person, which is coincident with the visual presence of a social stimulus, and therefore facilitate brain reactions to faces.

Notably, this effect was independent of emotional expressions. Thus, our hypothesis of stronger processing of the fear-related stimulus anger revealed no support. Although Hummer and McClintock (2009) also revealed no differentiation of androstadienone effects between angry and happy faces in a dot probe task, they revealed no effect on attention to neutral faces. Comparing the latter with the current study one has to take methodological differences into account. Whereas in the dot probe task emotional and neutral faces compete for attentional resources simultaneously, for EEG measurement emotional and neutral stimuli were presented consecutively. Furthermore, in the current study attention was explicitly directed towards the displayed emotion, whereas in the dot probe task attention was directed towards the dot appearing after subliminal presented faces. Therefore, it is still unclear why androstadienone did not specifically enhance the processing of anger. Whether stimulus features, the task at hand or contextual and individual characteristics may be responsible remains to be clarified.

No further visual ERP components were modulated by androstadienone exposure: as the P100 was not affected, androstadienone seems not to influence initial, automatic brain processes. Neither did androstadienone affect the rapid global facial encoding process reflected by the N170. Nevertheless, data indicated appropriate data acquisition. In contrast to P100, the face specific component was influenced by the emotional expression. Happy and angry compared to neutral faces elicited a more pronounced peak in the right hemisphere. This is in line with former research also using schematic faces (Kolassa et al., 2007; Kolassa *et al.*, 2009). Moreover, happy faces elicited an LPP compared to neutral faces as has been found previously (Mühlberger et al., 2009; Sato et al., 2001). However, in contrast to earlier literature (Mühlberger et al., 2009; Sato et al., 2001; Schupp et al., 2004) anger did not elicit an LPP and anger and happiness did not elicit an EPN compared to neutral faces. This inconsistency might be due to different stimulus material. Former studies presented real human faces or avatars, whereas the current study presented cartoon expressions. Kolassa et al. (2007), to our knowledge the first who analyzed the LPP in response schematic emotional faces, also did not find enhanced amplitudes in response to emotional compared to neutral expressions. Authors even found the opposite: an LPP to neutral compared to emotional cartoon faces. They suggest that their neutral face with its round eyes rather depict threat than neutrality and therefore might lead to an oddball effect. Therefore, we replaced these round eyes with "happy eyes" in our neutral face to prevent this pop out effect. Although subjective ratings elicited expected valence and arousal results, emotional, especially the angry faces still did not elicit LPP or EPN compared to the neutral expression in the control condition. This proposes that additional facial features apart from the emotional expression are crucial for eliciting LPP and EPN. However, as EEG studies with schematic faces are rare, this argument has to be tested in future experiments.

Another reason for missing emotional EPN might be the different data acquisition. Compared to former studies the occipital electrodes PO9, PO10, O9 and O10 were missing in our equipment. Moreover, the combination of olfactory and visual stimulation in general may alter central nervous stimulus processing. As this was the first experiment investigating the influence of an odor on central nervous visual processing, the impact of the control odor apart from androstadienone is not known.

Nevertheless, cartoons elicited expected valence and arousal ratings suggesting appropriate stimulus selection in the control group. The happy face was rated as more pleasant than angry and neutral faces, whereas the angry face was rated as less pleasant than happy and neutral faces. Also, happiness and anger were rated as similar intense and both emotions as more intense, i.e. more arousing, than the neutral face. Subjective ratings were not affected by androstadienone inhalation, which is in line with Lundström et al. (2003), suggesting a rather subconscious effect mechanism of the endogenous odor. Moreover, participants rated the perceived compounds as iso-intense and similar in pleasantness and familiarity, ensuring that odor quality is not responsible for the presented effect.

To be mentioned, in contrast to Kolassa et al.'s (2007) findings, in the current control condition the P300 was not modulated by expressed emotions. Indeed, both studies used similar stimulus material and the same emotion identification task, but higher statistical power in Kolassa's study by including many more men (29) and also women (27) may account for this difference. However, inconsistent results and the fact that the EPN component has never been investigated with schematic facial stimuli ask for further studies investigating brain reactions to schematic emotional faces.

To conclude, data showed stronger brain reactions, indicated by larger P300 amplitudes, towards faces while men were smelling androstadienone. This suggests that the body odors compound might improved the allocation of attentional resources towards these social stimuli. This supports the notion that the endogenous odor affects higher cognitive functions and is able to attune the mind towards presented visual stimuli. That the subjective perception of presented faces was not affected imply subconscious effect mechanisms of a non-detectable androstadienone concentration.

6 General discussion

6.1 *Androstadienone and explicit measures*

None of this thesis' studies revealed an influence of androstadienone on explicit measures, i.e. subjective ratings of faces or other affective stimuli. This might be due to the laboratory setting, as the only study proving such modulations investigated androstadienone effects in a speed-dating field experiment (Saxton et al., 2008). Women exposed to androstadienone compared to women exposed to a control odor rated their male speed dating partner as more attractive. This event provided much more external stimulation, which presumably distracted the women from the person who had to be judged. Based on this background distraction androstadienone might have been able to direct the women´s attention back to the social stimulus and as a consequence cause improved attractiveness ratings. In contrast, in a laboratory setting attention is already focused on presented social stimuli and therefore androstadienone might not be able to further enhance the attentional focus on the rating task. This might have been the case in this thesis' studies as well as in other laboratory experiments not detecting rating effects of androstadienone (Hummer & McClintock, 2009; Lundström & Olsson, 2005).

Alternatively, it has been shown that subliminal smells, rather than consciously detectable odors, are necessary to affect social preferences (Li, Moallem, Paller, & Gottfried, 2007). Neutral faces were rated as less likeable following a subconscious presentation of an unpleasant odor. However, this effect decreased with an increase of the ability to consciously detect the odor. In our studies, although presented on a subliminal concentration, we masked androstadienone with a consciously noticeable smell of clove. Therefore, existing effects of androstadienone might have been superseded by the supra-threshold aroma of clove oil.

If either the first or the latter explanation or even both may be responsible for negative results in this thesis has to be clarified in the future.

6.2 Effect mechanisms of androstadienone

Jacob and colleagues (2001a) found androstadienone induced changes in attention and vision related cortical areas, while participants had to complete a visual tracking task. This suggests that androstadienone acts via higher cognitive mechanisms. These mechanisms were sought to be tapped in study II and III of this thesis by using electroencephalography. Explorative results of study II tentatively proposed that androstadienone influences cognitive processes. Study III indeed showed androstadienone modulated later brain reactions where attentional processes play a crucial role. The endogenous odor elicited larger P300 amplitudes in response to faces, which indicates a facilitated central nervous face processing by enhancing attention towards these visual stimuli, at least in men and in respect to schematic cartoon faces. In line, study I revealed an androstadienone related enhancement of approach tendencies in men towards faces. Also this phenomenon may be explained by an enhanced attentional allocation. Androstadienone may signal the company of a potential interaction partner in addition to the visual presence. This congruency of information modalities might intensify attention and then result in a more positive appraisal and approach tendency. Also the revealed androstadienone related enhancement of attention based reactions from study I and results of former studies (Hummer & McClintock, 2009; Lundström et al., 2003) support the notion of androstadienone to have impact on our brain by modulating attentional processes.

EEG results and the improved approach score revealed from this thesis suggested that androstadienone enhances attention to faces independent of their expressions. But Hummer and McClintock (2009) rather suggest an androstadienone related enhancement of attention to emotional but not neutral information. In contrast, motoric response data from study I supposed an emotion specific modulation, namely an enhanced processing of anger but not of happiness by androstadienone. So, does androstadienone specifically interact with threatening stimuli or general emotionally affective information or does androstadienone act totally independent of the emotional content?

EEG and behavior studies in the current thesis demanded the same cognitive load, namely the identification of expressed emotion and following motoric reaction. Therefore, in both studies similar brain resources were engaged. However, the EEG trial provided a higher level of difficulty. The number of stimuli and thus response alternatives differed between studies. The EEG study provided three choices (happy, angry and neutral) while in the motoric reaction task there were only two possibilities (happy, angry). Furthermore, the third expression in the EEG study was non emotional, which is an ambiguous signal, being in some way more difficult to identify compared to happy and angry expressions. As the EEG study suggested an emotion independent and results of speeded reactions suggested an emotion specific androstadienone effect, one might conclude that the impact of androstadienone is less sophisticated if higher cognitive load is demanded by the task at hand. In other words, severity might be a crucial variable in androstadienone research.

However, the approach score, assessed with the same demand as speeded arm movements, suggested a rather emotional independent androstadienone effect, which challenges this explanation. Moreover, a comparison with Hummer´s and McClintock´s (2009) study is difficult, because of different methods. Whereas in the dot probe task (Hummer & McClintock, 2009) emotional and neutral faces simultaneously compete for attentional resources, in EEG and behavior measurements stimuli were presented in a row. Furthermore, the requested test of studies varied. In current thesis attention was clearly directed towards the displayed emotion, whereas in the dot probe task attention was directed towards the dot appearing after subliminal presented faces.

In sum, behavioral and central nervous data corroborates earlier suggestions that androstadienone yields effects on cognitive functions via attentional mechanisms. However, the emotional dependence of androstadienone effects could not be determined. Therefore, experimental settings and statistical analyses should be carefully selected to prevent crucial impact on the specificity of androstadienone´s effects.

6.3 Contextual variables

It has been stated that the meaning of specific circumstances defines whether a person responds to androstadienone or not (Jacob & McClintock, 2000). Already the animal literature demonstrated that a valid context is fundamental for pheromonal effects (cf. Beauchamp, Doty, Moultan, & Mugford, 1976; cf. Beauchamp, Doty, Moulton, & Mugford, 1979). In previous human studies (Bensafi, Brown et al., 2004; Bensafi et al., 2003) the emotional context, i.e. the induced mood of participants by arousing films, was claimed to be essential. Villemure and Bushnell (2007) called this assumption into question by suggesting that "the appropriate context is not so much the arousing nature of the films but rather the involvement of human beings (real or on screen) that is the key" (p. 190). Four different picture categories in study II, i.e. real faces, pictures with couples, pictures with social and non-social scenes, each including three different valence categories, i.e. positive, negative and neutral, should verify whether people (on screen) are necessary to elicit androstadienone effects or not. Explorative results suggested that androstadienone only affects cortical reactions to faces but not to social or non-social scenes. This implies that not just the presence of people on screen but rather the simulation of an immediate interaction partner by directly facing the observer is important in an effective context. Study III showing androstadienone modulated central nervous face processing encourage this point of view.

Alternatively, in contrast to the context displayed on pictures rather social circumstances might be relevant for tapping androstadienone related effects. Some studies define the gender of the experimenter as the relevant social context. For example, women´s responsiveness to androstadienone increased significantly, if tested by a man compared to a women (Bensafi, Brown et al., 2004; Jacob et al., 2002; Jacob, Hayreh et al., 2001; Lundström & Olsson, 2005; but see Villemure & Bushnell, 2007). In line are results of the current thesis, where male participants tested by a female experimenter showed a changed brain response and an enhanced approach tendency. However, also women in this theses all tested by a female experimenter, showed reactions to androstadienone: in study I, women reacted with similar response acceleration as men to androstadienone; in study II women showed a maintained cortisol level after inhaling androstadienone. These results rather challenge the assumption that an experimenter of the opposite sex is an essential factor for androstadienone related effects. Still, conclusions from this thesis remain cautious, because only a one female, but no male experimenter, tested the participants. Also influences of her specific personality cannot be ruled out.

Taken together, the context plays a crucial role for androstadienone effects. However, a context consists of many external variables like social interaction, daytime, task, odor application, stimulus material, additional measurements and internal variables like mood, experience or motivation, which results in countless combinations. To control for most of these variables and to provide a correct combination is highly challenging in a laboratory setting and even more in the field. Still, this interesting topic has to be further systematically investigated using different stimulus types best in everyday life situations.

6.4 Are androstadienone concentrations ecologically valid?

All effects in this thesis were modulated by minute amounts of androstadienone. The 250 µM concentration of androstadienone used was chosen to assure comparability with most previous experiments (e.g. Hummer & McClintock, 2009). However, it is understandable to argue that this concentration, which is circa 500 times higher than the natural occurrence of 17.9 pmol/cm², is too high to be ecologically valid. The application of supra-physiological levels of androstadienone may even result in absorption into the blood stream, where also pharmacological effects cannot be excluded.

One study investigated effects of two different androstadienone concentrations, 250 µM and 6250 µM, on mood and psychophysiology (Bensafi, Tsutsui, Khan, Levenson, & Sobel, 2004). Indeed, they detected a dose dependent effect. Both, mood and physiology were only affected by the higher stimulus concentration, which is however contradictory to several experiments using the lower concentration (Hummer & McClintock, 2009; Jacob et al., 2002; Jacob, Hayreh et al., 2001; Jacob, Kinnunen et al., 2001; Jacob & McClintock, 2000; Lundström et al., 2003; Lundström & Olsson, 2005). A pilot study testing androstadienone concentrations closer to natural occurrence (1 µM) did not detect effects on the autonomic nervous system or psychological variables (Lundström, Olsson, & Larsson, 2000). This indeed challenges the common practice of using a concentration, which is circa 500 times higher (or even more when used in crystalline form) than the concentration naturally produced by humans. Therefore, Jacob and McClintock (2000) tested the definite amount, which is delivered by a Q-tip applying 130 µl of a 250 µM androstadienone solution upon the upper lip, which is a common application method. They measured approximately 9 nm of androstadienone remaining on the participant´s skin. Therefore, the compound concentration finally reaching the nasal cavity seems to be very close to natural levels. This application method was adopted in study II of this thesis. In study I and III, androstadienone was applied via a constant air flow olfactometer (Lundström et al., 2010), delivering two liters of air per minute directly into the nasal cavity. This ensured a constant application of air from the headspace of a 250 µM compound solution, which better mimics the natural way of sensing androstadienone. Although we do not know exactly how much of the compound finally reaches the nose over a 10 minute time period, we suggest a much lower concentration than that of the liquid. However, this must be seen as first attempt to isolate androstadienone effects by a more natural way of application.

6.5 Androstadienone and the pheromone concept

6.5.1 The pheromone concept in question

Several aspects of the pheromone concept has been highly criticized (Doty, 2003b, 2010). First of all, with uncountable definitions it is almost impossible to prove or disprove, whether a chemical substance is a pheromone or not. Moreover, not one single isolated compound fulfills the pheromonal demands described in most definitions, innateness, species specificity and uniqueness of released response. Moreover, biologically active odors are typically not single pure chemicals, but rather complex mixtures composed of hundreds of volatile substances (Maarse, 1991). In addition complex animals, especially humans, have a highly evolved brain. Top down processes filter the noisy environment for significant information, which changes with the surrounding context over time (Wilson & Stevenson, 2006). Multiple brain regions take part in the processing of olfactory information. Olfactory information is projected from the olfactory bulb to the primary olfactory cortex, with the olfactory regions including the hippocampus, hypothalamus, thalamus, orbitofrontal and insular cortices and cingulate gyrus (Zatorre, Jonesgotman, Evans, & Meyer, 1992). Therefore, the meaning of an incoming stimulus is influenced by higher brain levels, where also past experiences are encoded. Learning, which actually contradicts the pheromone concept of innateness, is therefore a doubtless aspect of reactions to chemical signals, at least in humans. Additionally, the division into two classes of stimuli, pheromonal and non-pheromonal, is problematic "since mutually exclusive categories cannot share attributes or features and preclude the existence of multiple classes or continua. Such dichotomies [...] limit the range of possible options, forcing adherents to fit any number of phenomena into one or the other class. The attempts to define pheromones on the basis of such bipartite categories as innate versus learned, single versus multiple, conspecific versus heterospecific, olfactory versus vomeronasal, volatile versus nonvolatile and hypothalamic versus non-hypothalamic seem doomed from the start" (p. 188) (Doty, 2010). Therefore, scientists were forced to modify the term pheromone. They added descriptors and created new classes like signaling pheromones and releasing pheromones or in humans the modulator pheromones.

Moreover, the pheromone concept neglects the multimodality of communication. Instead, it implies the existence of single substances eliciting unique responses without taking other senses sight, hearing, touch and taste, into account. Therefore, using the insect based term "pheromonal" to report behavioral or psychological reactions in humans is an oversimplification of complex phenomena.

The belief that human pheromones exist comes from body odors studies. Some of them have been criticized regarding the methodological assessment and statistics aspects. One study suggested that axillary secretions convey specific social information, namely the sex of another person (Russell, 1976). However, also a higher intensity of male compared to female sweat may cause differences in sweat odor between sexes. Support for this hypothesis comes from a study in which only female sweat was presented (Doty, Orndorff, Leyden, & Kligman, 1978). The author found a strong correlation between intensity ratings of the presented sweat samples and the likeliness to assign it to the male category. Female participants recognized the strongest odors to be male, whereas the weakest one was believed to be female. In this context it is noteworthy that men have larger apocrine glands than women and that the gland size correlates with odor intensity, which make intensity variation between sexes even more likely. Although, odor variability is partly caused by genetic factors also environmental variables, like e.g. diet, are important aspects, which can significantly alter sweat smell. For example, body odors sampled from meat dieters were rated as less pleasant, less attractive and more intense than samples from non-meat dieters. This stresses the individual food pattern as an important factor in odor variability (Havlicek & Lenochova, 2006). As men consume more meat than women (Shiferaw et al., 2008), this may also account for sex differences in sweat odor.

Another study reported an influence of axillary odors on participant´s mood (Chen & Haviland-Jones, 1999). This study presented sweat samples from different age and sex groups to 300 volunteers. Only participants exposed to underarm odors of older adults reported a reduction in negative mood. However, is has been criticized that the used mood scale was not appropriate to assess mood changes, but rather to assess frequencies in experiencing these feelings. Moreover, a placebo controlled condition was not included in their statistical analyses (Black, 2001).

The study by Thorne et al. (2002) reported a positive modulation of sexual attractiveness ratings of male photographs and short written narratives when women were exposed to male axillary secretions. Notably, taking hormonal contraceptives improved those ratings. This seems questionable because the effect might also have been due to mood changes caused by oral contraceptives (Oinonen & Mazmanian, 2002). Oral contraceptive user´s mood is known to be more stable and less depressive than that of spontaneously ovulating women. Moreover, effects were compared to a no-odor control condition. Although authors reported that presented body odors samples did not emit a noticeable odor, still other odorants might induce same effects.

The *Nature* paper by Stern and McClintock (1998), reported that odorless compounds from female armpits changed the length of the menstrual cycle phases of recipient women. Timing of ovulation was manipulated depending on the menstrual cycle phase, in which donor sweat was collected. Authors concluded the proved existence of "two distinct pheromones" (p. 178) in female axillary secretions. Although, more studies reported a synchronization of menses induced by female axillary secretions (Preti, Cutler, Garcia, Huggins, & Lawley, 1986; Russell et al., 1980), several authors highly criticized those results. It was shown that applying the same statistics with same means and standard deviations as in Stern´s and McClintock´s study on a random data set results become also significant (Schank, 2006). Whitten (1999) stated that excluding the two outliers favorable to the model in Stern´s and McClintock´s report would eliminate significant changes in cycle lengths. Furthermore, he criticized the use of only one single cycle as baseline length, which has "no within-subject variance and the irregular statistical maneuvers of converting all 20 observations to zero masks any between-subject variance" (p. 232) (Whitten, 1999). In Russell´s study (1980), also reporting menstrual synchronization after body odors exposure, neither participants nor the experimenter were blind to the purpose of the study, which may confound reported results. Generally, a changed cycle length over time may likely be due to natural variation in women´s menstrual cycle (Strassmann, 1999). About half of tested women tend to synchronize cycles incidentally, when observed only for a few cycles (Wilson, 1987). Moreover, to control for cycle onsets is not sufficient to determine an absolute difference between women, as cycles vary in length.

In sum, reported pheromone-mediated phenomena elicited by human axillary secretions remain questionable on the basis of statistical and methodological grounds. Additionally, a substance isolated from human sweat, which could have been called a pheromone, because of making women discernable from men or synchronizing menstrual cycles, is still missing.

Still, as reviewed in this thesis, the single sweat compound androstadienone has been found to be indeed affective in humans. But also these studies have to be interpreted with care. Brain imaging studies reporting a sex-specific androstadienone influence on hypothalamus activation exclusively used PET measurement (Berglund et al., 2008; Frasnelli et al., 2008; Jacob, Kinnunen et al., 2001; Savic, Berglund, & Lindstrom, 2006; Savic et al., 2005). However, PET scanning with a limited spatial resolution might hinder conclusions about specific activations of hypothalamic nuclei. Also critical is the use of non-matched common odors in Savic and colleagues' studies. Those stimuli were different in structure and intensity compared to androstadienone. Differences in brain activation might therefore also result from varying combinations of activated olfactory cells. Moreover, without any behavioral or endocrinological correlates the implications of Savic et al.'s findings are vague.

A general drawback in androstadienone research is the use of different measures, compound application methods and contexts, which prevents direct comparison between studies. Moreover, large sample sizes and statistical methods to reduce variance are necessary to detect androstadienone effects, which indicate them to be small and instable.

An also crucial point is that androstadienone never occurs as a single compound in nature. Rather the composition of the smell of an individual, than one particular substance, may contribute to olfactory and social information. Moreover, the amount of androstadienone found in sweat is low and highly variable and some men do not secrete it at all (Nixon et al., 1988). Nevertheless, androstadienone was found to affect psycho-physiological and behavioral reactions in humans, which makes it a promising candidate for an active human substance.

6.5.2 Is androstadienone a human pheromone?

One general aim of this thesis is to provide further data clarifying, whether the steroidal compound androstadienone should be labeled a human pheromone. However, aforementioned problems arising from the pheromone concept, call for a careful discussion of pheromonal characteristics and effects of androstadienone. Does androstadienone really fit the pheromone concept which suggests a single, externally secreted compound, eliciting a unique response in conspecifics, which is rather innate than learned?

Derived from testosterone androstadienone indeed fits the first requirement of the pheromone concept as a single compound. Although, it does not contribute much to the characteristic smell of individual body odors, which largely consists of unsaturated acids (Zeng et al., 1992; Zeng et al., 1991), it is definitely present in human sweat (Gower et al., 1994). Thus also the second criterion, an externally secreted compound, is met by androstadienone. It is secreted by one individual to the outside and therefore it is likely to be transferred to another person. However, the hypothesis that a transfer of social information between humans with androstadienone takes place remains questionable. The third requirement of the pheromone concept, species specificity, is not met, because it is also found in pigs. Also behavioral effects caused by androstadienone are extremely different between pigs and humans, which contradicts the criterion of one specific response to androstadienone. Moreover, several studies reported influence of androstadienone on higher brain levels like the visual and parietal cortices, the fusiform and superior temporal gyri, the hypothalamus, the prefrontal cortex, the cingulate cortex and the amygdala (Gulyas et al., 2004; Jacob, Kinnunen et al., 2001). With the involvement of areas related to attentional mechanisms or social cognition, responses to androstadienone do not confirm the forth necessity of the pheromone concept: innateness and unchangeable responses.

Another important question is, whether effects attributed to androstadienone are unique, the fifths criterion for a pheromone, or whether other odorants may have similar influence on human physiology and psychology. Androstadienone has been repeatedly shown to modulate sympathetic nervous reactions in men and women. It increases physiological arousal in women, indicated by an increase in skin conductance, heart rate, blood pressure and respiration rate and a decrease in skin temperature (Bensafi et al., 2003; Jacob, Hayreh et al., 2001; Lundström & Olsson, 2005; Wyart et al., 2007). However, that also plant aromas can induce changes in human emotions, behavior and autonomic nervous system function has been reported previously (Moss, Hewitt, Moss, & Wesnes, 2008). Grapefruit oil, pepper oil, estragon oil and fennel oil are able to increase sympathetic nervous system functions in young women (cf. Doty, 2010, p. 164). Furthermore, androstadienone was shown to worsen memory performance in women but not in men related to a sad film clip (Bensafi, Brown et al., 2004). Also ylang-ylang was reported to impair the quality of memory in students (Moss et al.,

2008). Another repeatedly reported effect of androstadienone is the improvement of female's mood (Jacob, Hayreh et al., 2001; Jacob & McClintock, 2000; Lundström & Olsson, 2005; Villemure & Bushnell, 2007; Wyart et al., 2007). In accordance, also pleasant common odors are able increase positive mood, whereas unpleasant odors rather induce a negative affective state (Ehrlichman & Bastone, 1992). Lemon oil, for example, enhances positive mood relative to water and lavender oil (Kiecolt-Glaser *et al.*, 2008). Brain imaging PET studies demonstrated that androstadienone, compared to air, estratetraenol, lavender oil, cedar oil, eugenol and butanol activates the hypothalamus (Berglund et al., 2006; Frasnelli et al., 2008; Savic et al., 2001; Savic et al., 2005). In contrast, common control odors activated the amygdala, piriform, orbitofrontal and insular cortices, suggesting indeed specific characteristics of the steroid used in those studies. However, one fMRI study did also detect a hypothalamus activation by lavender (Wang, Eslinger, Smith, & Yang, 2005). It is important to note that effects of common odors on physiology and mood but also on brain activity were not dependent on participant's sex or sexual orientations, which is the case for effects of androstadienone. In men compared to women androstadienone induces a rather negative mood and decreases physiological arousal (Bensafi, Brown et al., 2004; Bensafi et al., 2003; Jacob & McClintock, 2000; Villemure & Bushnell, 2007). Also, the only study directly comparing effects of androstadienone with those of a pleasant common odor reported a sex-dimorphism for androstadienone (Villemure & Bushnell, 2007). Whereas the common pleasant odor improved mood in all participants androstadienone did so only in women. Interestingly androstadienone related hypothalamus responses were found in heterosexual women and homosexual men but not in heterosexual men or homosexual women (Berglund et al., 2006; Savic et al., 2006). Such a distinctive pattern was not demonstrated when presenting environmental odors. Also, several effects of androstadienone have not been replicated with common odors. Villemure and Bushnell (2007) reported an increased rating of pain intensity in women exposed to androstadienone. In contrast, Kiecolt-Glaser and colleagues (2008) did neither find a reliably altered rating of pain by lemon oil or lavender oil following a cold pressor stress test, nor did those odors reliably alter salivary cortisol levels, which was found for androstadienone (Wyart et al., 2007).

Also in the current thesis androstadienone modulated obvious behavior in terms of speeded muscle activation and approach tendencies and altered the central nervous processing of emotional faces. To my knowledge, common odors have not yet been shown to modulate these reactions in humans. But it has to be mentioned that so far only a limited number of common odorants have been studied and rarely both sexes were tested within the same experimental setting. Moreover, different paradigms and stimulus presentation methods hinder the comparison of results across studies. Most important, studies directly comparing effects of androstadienone and environmental odors with similar structure and intensities are rare but necessary to definitely conclude unique effects of androstadienone.

Taken together, androstadienone is an isolated compound and is produced and externally secreted by one individual. Although, common odors have been shown to modulate arousal, mood and brain responses, androstadienone´s sexual specificity on these measurements as well as its effectiveness in subthreshold amounts on motoric reactions and behavioral tendencies have not been demonstrated for common odorants. Although, this still suggests a singularity of the androstadienone function, it is not a species specific substance and with the engagement of higher cognitive functions androstadienone fails the classical assumption of eliciting innate responses. Therefore, androstadienone may not be claimed a human pheromone.

6.5.3 Is androstadienone a human modulator pheromone?

As has been demonstrated, the classical pheromone definition criteria, derived from insect research, are too strict when taking human complexity into account. Especially, the development of the human cortex makes a single stimulus unlikely to elicit an immediate reliable response. It might rather alter an individual´s way of responding without necessarily evoking an immediate observable answer (Preti & Wysocki, 1999), as complex integration of all senses leads to human reactions. Also development and experiences are vital factors for a complex reaction. Moreover, communication is not species specific. It is also influenced e.g. by inter-specific interactions between prey and predator, which has been under evolutionary pressure (Doty, 2010).

Accordingly, McClintock (2003) published a definition matching characteristics of human endogenous compounds in a better way. She assumed that a human chemosignal, rather than initiating a reaction, modulates ongoing behavior or psychological reactions to a particular context, and called it modulator pheromone. She assumed that alterations caused by this chemosignal might be driven by changes in stimulus sensitivity, salience or sensory-motor integration. According to that definition, the current thesis demonstrates that androstadienone modulates ongoing behavior and psycho-physiological responses, namely speeded motor reactions and the processing of visual stimuli via modulation of attention related networks. Additionally, androstadienone might have altered the sensory-motor integration, i.e. the integration of visual input with motoric output found in study I, also meeting McClintock's definition. Via enhanced psychological attention towards anger androstadienone might have facilitated the stimulus encoding and therefore the appropriate behavioral reaction. Moreover, androstadienone might change stimulus sensitivity resulting in a modulated evaluation of stimuli, which then might have improved approach tendencies measured in study I.

It is important to note that androstadienone's subthreshold concentrations prevented effects to be due to hedonic factors and were exclusively elicited by compound characteristics. Together with previous studies representing influences of androstadienone on basic psychological and physiological measures, current results agree with the notion that androstadienone acts like a human modulator pheromone. This however, remains tentative until future research replicates these findings, first, in comparison with other environmental odors, second, in ecologically valid environments and third, with concentrations similar to natural occurrence.

6.6 Concluding remarks

Although, previous research suggests that androstadienone meets especially the criteria of a human modulator pheromone, I suppose that it is not necessary to work with the term "pheromone". Even more important is that androstadienone has indeed the ability to influence human's autonomous nervous system, psychological state, as well as brain reactions and obvious behavior in a specific way. A crucial aspect is that it is affective even in minute amounts not perceivable as an odor, influencing us on a subconscious level. These facts and that androstadienone is permanently present during social interactions, makes it a promising candidate for an active social chemosignal. This thesis provides a further step in characterizing modulator effects of androstadienone influencing higher cognitive functions. For the first time a significant androstadienone related influence on the central nervous processing of visual cues was proven. Moreover, it was shown that androstadienone is able to modulate behavior and motivational tendencies in humans. This is promising with regard to the ability of androstadienone to guide our behavior in everyday life situations.

Certainly results seek for replication especially in an ecologically valid context. As field experiments are highly complex with uncountable factors difficult to control, I suggest for future studies an intermediate step in a more controllable, but still ecologically suitable context – the virtual reality.

7 References

Albrecht, J., Boesveldt, S., Gordon, A. R., Alden, E. C., Hernandez, M. F., & Lundström, J. N. (submitted). Test-retest reliability of the 40-item Monell Extended Sniffin' Sticks Identification Test (MONEX-40).

Amaral, D. G., Price, J. L., Pitkanen, A., & Carmichael, S. T. (1992). Anatomical organization of the primate amygdaloid complex. In J. Aggleton (Ed.), *The amygdala: neurobiological aspects of emotion, memory and mental dysfunction* (pp. 1-66). New York: Wiley.

Beauchamp, G. K., Doty, R. L., Moultan, D. G., & Mugford, R. A. (1976). The pheromone concept in mammalian chemical communication: a critique. In R. L. Doty (Ed.), *Mammalian olfaction, reproductive processes and behavior* (pp. 143-160). New York: Academic press.

Beauchamp, G. K., Doty, R. L., Moulton, D. G., & Mugford, R. A. (1979). Defense of the term pheromone - reply. *Journal Of Chemical Ecology, 5*(2), 301-305.

Beauchamp, G. K. & Yamazaki, K. (2003). Chemical signaling in mice. *Biochemical Society Transactions, 31*, 147-151.

Bensafi, M., Brown, W. M., Khan, R., Levenson, B., & Sobel, N. (2004). Sniffing human sex-steroid derived compounds modulates mood, memory and autonomic nervous system function in specific behavioral contexts. *Behavioural Brain Research, 152*(1), 11-22.

Bensafi, M.Brown, W. M.Tsutsui, T.Mainland, J. D.Johnson, B. N.Bremner, E. A.Young, N., et al. (2003). Sex-steroid derived compounds induce sex-specific effects on autonomic nervous system function in humans. *Behavioral Neuroscience, 117*(6), 1125-1134.

Bensafi, M., Tsutsui, T., Khan, R., Levenson, R. W., & Sobel, N. (2004). Sniffing a human sex-steroid derived compound affects mood and autonomic arousal in a dose-dependent manner. *Psychoneuroendocrinology, 29*(10), 1290-1299.

Bentin, S., Allison, T., Puce, A., Perez, E., & McCarthy, G. (1996). Electrophysiological studies of face perception in humans. *Journal Of Cognitive Neuroscience, 8*(6), 551-565.

Benton, D. (1982). The influence of androstenol - a putative human pheromone - on mood throughout the menstrual cycle. *Biological Psychology, 15*, 249-256.

Benton, D. & Wastell, V. (1986). Effects of androstenol on human sexual arousal. *Biological Psychiatry, 22*, 141-147.

Berglund, H., Lindström, P., Dhejne-Helmy, C., & Savic, I. (2008). Male-to-female transsexuals show sex-atypical hypothalamus activation when smelling odorous steroids. *Cerebral Cortex, 18*(8), 1900-1908.

Berglund, H., Lindström, P., & Savic, I. (2006). Brain response to putative pheromones in lesbian women. *Proceedings Of The National Academy Of Sciences Of The United States Of America, 103*(21), 8269-8274.

Berliner, D. L., Monti-Bloch, L., Jennings-White, C., & Diaz-Sanchez, V. (1996). The functionality of the human vomeronasal organ (VNO): Evidence for steroid receptors. *Journal Of Steroid Biochemistry And Molecular Biology, 58*(3), 259-265.

Bethe, A. (1932). Vernachlässigte Hormone. *Naturwissenschaften*, 177-181.

Black, S. L. (2001). Does smelling granny relieve depressive mood? Commentary on 'Rapid mood change and human odors'. *Biological Psychology, 55*(3), 215-218.

Boesveldt, S., Frasnelli, J., Gordon, A. R., & Lundström, J. (2010). The fish is bad: negative food odors elicit faster and more accurate reactions than other odors. *Biological Psychology, 84*, 313-317.

Boyle, J. A., Lundström, J. N., Knecht, M., Jones-Gotman, M., Schaal, B., & Hummel, T. (2006). On the trigeminal percept of androstenone and its implications on the rate of specific anosmia. *Journal Of Neurobiology, 66*(13), 1501-1510.

Bremner, E. A., Mainland, J. D., Khan, R. M., & Sobel, N. (2003). The prevalence of androstenone anosmia. *Chemical Senses, 28*(5), 423-432.

Broca, P. (1888). *Mémoires d´Anthropologie (Vol. 5)*. Paris: Reinwald.

Brooksbank, B. W. L., Wilson, D. A. A., & MacSweeney, D. A. (1972). Fate of androsta-4,16-dien-3-one and the origin of 3-alpha-hydroxy-5alpha-androst-16-ene in man. *Journal Of Endocrinology, 52*(2), 239-251.

Bruce, H. M. (1959). Exteroceptive block to pregnancy in the mouse. *Nature, 184*(4680), 105-105.

Buck, L. B. (2000). The molecular architecture of odor and pheromone sensing in mammals. *Cell, 100*(6), 611-618.

Bullivant, S. B., Sellergren, S. A., Stern, K., Spencer, N. A., Jacob, S., Mennella, J. A., & McClintock, M. K. (2004). Women's sexual experience during the menstrual cycle: identification of the sexual phase by noninvasive measurement of luteinizing hormone. *Journal Of Sex Research, 41*(1), 82-93.

Cacioppo, J. T., Priester, J. R., & Berntson, G. G. (1993). Rudimentary determinants of attitudes. II: Arm flexion and extension have differential effects on attitudes. *Journal of Personality and Social Psychology, 65*(1), 5-17.

Campbell, D. T. (1963). Social attitudes and other acquired behavioral dispositions. In S. Koch (Ed.), *Psychology: A study of a science* (Vol. 6, pp. 94-172). New York: McGraw-Hill.

Chen, D. & Haviland-Jones, J. (1999). Rapid mood change and human odors. *Physiology & Behavior, 68*(1-2), 241-250.

Chen, D. & Haviland-Jones, J. (2000). Human olfactory communication of emotion. *Perceptual And Motor Skills, 91*(3), 771-781.

Chen, D., Katdare, A., & Lucas, N. (2006). Chemosignals of fear enhance cognitive performance in humans. *Chemical Senses, 31*(5), 415-423.

Chen, M. & Bargh, J. A. (1999). Consequences of automatic evaluation: immediate behavioral predispositions to approach or avoid the stimulus. *Personality and Social Psychology Bulletin, 25*(2), 215-224.

Claus, R. & Alsing, W. (1976). Occurrence of 5-alpha-androst-16-en-3-one, a boar pheromone in man and its relationship to testosterone. *Journal Of Endocrinology, 68*(3), 483-484.

Cornwell, R. E.Boothroyd, L.Burt, D. M.Feinberg, D. R.Jones, B. C.Little, A. C.Pitman, R., et al. (2004). Concordant preferences for opposite-sex signals? Human pheromones and facial characteristics. *Proceedings Of The Royal Society Of London Series B-Biological Sciences, 271*(1539), 635-640.

Cowley, J. J., Johnson, A. L., & Brooksbank, B. W. L. (1977). Effect of two odorous compounds on performance in an assessment-of-people test. *Psychoneuroendocrinology, 2*(2), 159-172.

Cuthbert, B. N., Schupp, H. T., Bradley, M. M., Birbaumer, N., & Lang, P. J. (2000). Brain potentials in affective picture processing: covariation with autonomic arousal and affective report. *Biological Psychology, 52*(2), 95-111.

Dorries, K. M., Schmidt, H. J., Beauchamp, G. K., & Wysocki, C. J. (1989). Changes in sensitivity to the odor of androstenone during adolescence. *Developmental Psychobiology, 22*(5), 423-435.

Doty, R. L. (1981). Olfactory communication in humans. *Chemical Senses, 6*(4), 351-376.

Doty, R. L. (2003a). *Handbook of olfaction and gustation* (2 ed.). New York: Informa Healthcare.

Doty, R. L. (2003b). Mammalian pheromones: fact or fantasy? In R. L. Doty (Ed.), *Handbook of olfaction and gustation* (2 ed., pp. 345-383). New York: Informa Healthcare.

Doty, R. L. (2010). *The great pheromone myth*. Baltimore: Johns Hopkins University Press.

Doty, R. L., Orndorff, M. M., Leyden, J., & Kligman, A. (1978). Communication of gender from human axillary odors - relationship to perceived intensity and hedonicity. *Behavioral Biology, 23*(3), 373-380.

Eastwood, J. D., Smilek, D., & Merikle, P. M. (2001). Differential attentional guidance by unattended faces expressing positive and negative emotion. *Perception & Psychophysics, 63*(6), 1004-1013.

Ehrlichman, H. & Bastone, L. (1992). The use of odour in the study of emotion. In S. Toller & G. H. Dodd (Eds.), *Fragrance. The psychology and biology of perfume* (pp. 143-159). London: Elsevier Applied Sciences.

Eimer, M. (2000). The face-specific N170 component reflects late stages in the structural encoding of faces. *Neuroreport, 11*(10), 2319-2324.

Eimer, M. & Holmes, A. (2002). An ERP study on the time course of emotional face processing. *Neuroreport, 13*(4), 427-431.

Eimer, M., Holmes, A., & McGlone, F. P. (2003). The role of spatial attention in the processing of facial expression: an ERP study of rapid brain responses to six basic emotions. *Cogn Affect Behav Neurosci., 3*(2), 97-110.

Eimer, M. & McCarthy, R. A. (1999). Prosopagnosia and structural encoding of faces: evidence from event-related potentials. *Neuroreport, 10*(2), 255-259.

Esteves, F., Parra, C., Dimberg, U., & Öhman, A. (1994). Nonconscious associative learning - pavlovian conditioning of skin-conductance responses to masked fear-relevant facial stimuli. *Psychophysiology, 31*(4), 375-385.

Filsinger, E. E., Braun, J. J., & Monte, W. C. (1985). An examination of the effects of putative pheromones on human judgments. *Ethology And Sociobiology, 6*(4), 227-236.

Fox, E., Russo, R., Bowles, R., & Dutton, K. (2001). Do threatening stimuli draw or hold visual attention in subclinical anxiety? *Journal of Experimental Psychology-General, 130*(4), 681-700.

Fox, E., Russo, R., & Dutton, K. (2002). Attentional bias for threat: Evidence for delayed disengagement from emotional faces. *Cognition & Emotion, 16*(3), 355-379.

Frasnelli, J., Lundström, J. N., Boyle, J. A., Katsarkas, A., & Jones-Gotman, M. (2008). Functional imaging after occlusion of the vomeronasal organ. *Chemical Senses, 33*(8), S89-S90.

Frey, M. C. M., Lundström, J. N., Weyers, P., Pauli, P., & Mühlberger, A. (submitted). Androstadienone modulates attention-based reactions towards angry faces in men and women.

Gaafar, H. A., Tantawy, A. A., Melis, A. A., Hennawy, D. M., & Shehata, H. M. (1998). The vomeronasal (Jacobson's) organ in adult humans: frequency of occurrence and enzymatic study. *Acta Oto-Laryngologica, 118*(3), 409-412.

Gower, D. B., Bird, S., Sharma, P., & House, F. R. (1985). Axillary 5-alpha-androst-16-en-3-one in men and women - relationships with olfactory acuity to odorous 16-androstenes. *Experientia, 41*(9), 1134-1136.

Gower, D. B., Holland, K. T., Mallet, A. I., Rennie, P. J., & Watkins, W. J. (1994). Comparison of 16-androstene steroid concentrations in sterile apocrine sweat and axillary secretions - interconversions of 16-androstenes by the axillary microflora - a mechanism for axillary odor production in man. *Journal Of Steroid Biochemistry And Molecular Biology, 48*(4), 409-418.

Grammer, K. (1993). 5-alpha-androst-16-en-3-alpha-on - a male pheromone - a brief report. *Ethology And Sociobiology, 14*(3), 201-207.

Gratton, G., Coles, M. G., & Donchin, E. (1983). A new method for off-line removal of ocular artifact. *Electroencephalography & Clinical Neurophysiology, 55*(4), 468-484.

Greenwald, A. G., McGhee, D. E., & Schwartz, J. L. K. (1998). Measuring individual differences in implicit cognition: the implicit association test. *Journal Of Personality And Social Psychology, 74*(6), 1464-1480.

Grosser, B. I., Monti-Bloch, L., Jennings-White, C., & Berliner, D. L. (2000). Behavioral and electrophysiological effects of androstadienone, a human pheromone. *Psychoneuroendocrinology, 25*(3), 289-299.

Gulyas, B., Keri, S., O'Sullivan, B. T., Decety, J., & Roland, P. E. (2004). The putative pheromone androstadienone activates cortical fields in the human brain related to social cognition. *Neurochemistry International, 44*(8), 595-600.

Hansen, C. H. & Hansen, R. D. (1988). Finding the face in the crowd: An anger superiority effect. *Journal of Personality and Social Psychology, 54*, 917-924.

Havlicek, J. & Lenochova, P. (2006). The effect of meat consumption on body odor attractiveness. *Chemical Senses, 31*(8), 747-752.

Havlicek, J. & Roberts, S. C. (2009). MHC-correlated mate choice in humans: A review. *Psychoneuroendocrinology, 34*(4), 497-512.

Haxby, J. V., Hoffman, E. A., & Gobbini, M. I. (2000). The distributed human neural system for face perception. *Trends Cogn. Sci., 4*, 223-233.

Herrick, C. J. (1924). *Neurological foundation of behavior.* New York: Holt.

Heuer, K., Rinck, M., & Becker, E. S. (2007). Avoidance of emotional facial expressions in social anxiety: The Approach-Avoidance Task. *Behaviour Research And Therapy, 45*(12), 2990-3001.

Holmes, A., Vuilleumier, P., & Eimer, M. (2003). The processing of emotional facial expression is gated by spatial attention: evidence from event-related brain potentials. *Cognitive Brain Research, 16*(2), 174.

Hummer, T. A. & McClintock, M. K. (2009). Putative human pheromone androstadienone attunes the mind specifically to emotional information. *Hormones And Behavior, 55*, 548-559.

Jackman, P. J. H. & Noble, W. C. (1983). Normal Axillary Skin Microflora In Various Populations. *Clinical And Experimental Dermatology, 8*(3), 259-268.

Jacob, S., Garcia, S., Hayreh, D., & McClintock, M. K. (2002). Psychological effects of musky compounds: comparison of androstadienone with androstenol and muscone. *Hormones And Behavior, 42*(3), 274-283.

Jacob, S., Hayreh, D. J. S., & McClintock, M. K. (2001). Context-dependent effects of steroid chemosignals on human physiology and mood. *Physiology & Behavior, 74*(1-2), 15-27.

Jacob, S., Kinnunen, L. H., Metz, J., Cooper, M., & McClintock, M. K. (2001). Sustained human chemosignal unconsciously alters brain function. *Neuroreport, 12*(11), 2391-2394.

Jacob, S. & McClintock, M. K. (2000). Psychological state and mood effects of steroidal chemosignals in women and men. *Hormones And Behavior, 37*(1), 57-78.

Johnson, A., Josephson, R., & Hawke, M. (1985). Clinical And Histological Evidence For The Presence Of The Vomeronasal (Jacobsons) Organ In Adult Humans. *Journal Of Otolaryngology, 14*(2), 71-79.

Johnston, R. E. (1998). Pheromones, the vomeronasal system, and communication - From hormonal responses to individual recognition *Olfaction And Taste Xii - An International Symposium* (Vol. 855, pp. 333-348).

Karlson, P. & Lüscher, M. (1959). Pheromones - New Term For A Class Of Biologically Active Substances. *Nature, 183*(4653), 55-56.

Keller, A., Zhuang, H. Y., Chi, Q. Y., Vosshall, L. B., & Matsunami, H. (2007). Genetic variation in a human odorant receptor alters odour perception. *Nature, 449*(7161), 468-U466.

Kiecolt-Glaser, J. K., Graham, J. E., Malarkey, W. B., Porter, K., Lemeshow, S., & Glaser, R. (2008). Olfactory influences on mood and autonomic, endocrine, and immune function. *Psychoneuroendocrinology, 33*(3), 328-339.

Kimchi, T., Xu, J., & Dulac, C. (2007). A functional circuit underlying male sexual behaviour in the female mouse brain. *Nature, 448*(7157), 1009-U1001.

Kinsey, A. C., Pomeroy, W. B., Martin, C. E., & Gebhard, P. H. (1953). *Sexual behaviour in the human female*. Philadelphia: Saunders.

Kirk-Smith, M. D. & Booth, D. A. (1980). Effects of androstenone on choice of location in others'presence. In v. d. S. H. (Ed.), *Olfaction and taste* (Vol. 7, pp. 389-392). London: IRL Press.

Kirk-Smith, M. D., Booth, D. A., Carroll, D., & Davies, P. (1978). Human social attitudes affected by androstenol. *Research Communications in Psychology, Psychiatry & Behavior, 3*(4), 379-384.

Knaapila, A.Tuorila, H.Silventoinen, K.Wright, M. J.Kyvik, K. O.Cherkas, L. F.Keskitalo, K., et al. (2008). Genetic and Environmental Contributions to Perceived Intensity and Pleasantness of Androstenone Odor: An International Twin Study. *Chemosensory Perception, 1*(1), 34-42.

Knecht, M., Kuhnau, D., Huttenbrink, K. B., Witt, M., & Hummel, T. (2001). Frequency and localization of the putative vomeronasal organ in humans in relation to age and gender. *Laryngoscope, 111*(3), 448-452.

Knecht, M., Lundström, J. N., Witt, M., Huttenbrink, K. B., Heilmann, S., & Hummel, T. (2003). Assessment of olfactory function and androstenone odor thresholds in humans with or without functional occlusion of the vomeronasal duct. *Behavioral Neuroscience, 117*(6), 1135-1141.

Kohl, J. V., Atzmueller, M., Fink, B., & Grammer, K. (2001). Human pheromones: Integrating neuroendocrinology and ethology. *Neuroendocrinology Letters, 22*(5), 309-321.

Kolassa, I.-T., Kolassa, S., Musial, F., & Miltner, W. H. R. (2007). Event-related potentials to schematic faces in social phobia. *Cognition and Emotion, 21*(8), 1721-1744.

Kolassa, S., Bergmann, S., Lauche, R., Dilger, S., Miltner, W. H. R., & Musial, F. (2009). Interpretive bias in social phobia: An ERP study with morphed emotional schematic faces. *Cognition and emotion, 23*(1), 69-95.

Kovacs, G.Gulyas, B.Savic, I.Perrett, D. I.Cornwell, R. E.Little, A. C.Jones, B. C., et al. (2004). Smelling human sex hormone-like compounds affects face gender judgment of men. *Neuroreport, 15*(8), 1275-1277.

Lang, P. J., Bradley, M. M., & Cuthbert, B. N. (1999). International affective picture system (IAPS): instruction manual and affective ratings. Universtity of Florida: The Center for Research in Psychophysiology.

Lang, P. J., Bradley, M.M., & Cuthbert, B.N. (1997). Motivated attention: Affect, activation, and action. *Attention and Orienting: Sensory and Motivational Processes*, 97-135.

Laska, M. & Freyer, D. (1997). Olfactory discrimination ability for aliphatic esters in squirrel monkeys and humans. *Chemical Senses, 22*(4), 457-465.

Laska, M., Genzel, D., & Wieser, A. (2005). The number of functional olfactory receptor genes and the relative size of olfactory brain structures are poor predictors of olfactory discrimination performance with enantiomers. *Chemical Senses, 30*(2), 171-175.

Laux, L., Glanzmann, P., Schaffner, P., & Spielberger, C. D. (1981). *Das State-Trait Angstinventar [The state–trait anxiety inventory]*. Weinheim: Beltz.

Li, W., Moallem, I., Paller, K. A., & Gottfried, J. A. (2007). Subliminal smells can guide social preferences. *Psychological Science, 18*(12), 1044-1049.

Linkenkaer-Hansen, K., Palva, J. M., Sams, M., Hietanen, J. K., Aronen, H. J., & Ilmoniemi, R. J. (1998). Face-selective processing in human extrastriate cortex around 120 ms after stimulus onset revealed by magneto- and electroencephalography. *Neuroscience Letters, 253*(3), 147-150.

Lubke, K., Schablitzky, S., & Pause, B. M. (2009). Male Sexual Orientation Affects Sensitivity to Androstenone. *Chemosensory Perception, 2*(3), 154-160.

Luck, S. J., Woodman, G. F., & Vogel, E. K. (2000). Event-related potential studies of attention. *Trends In Cognitive Sciences, 4*(11), 432-440.

Lundquist, D., Flykt, A., & Öhman, A. (1998). The Karolinska Directed Emotional Faces - KDEF. *CD ROM from Department of Clinical Neuroscience, Psychology section, Karolinska Institute, Stockholm, Sweden;* ISBN 91-630-7164-9.

Lundström, J. N. (2005). *Human Pheromones. Psychological and neurological modulation of a putative human pheromone.* Doctoral Thesis, Uppsala University, Uppsala.

Lundström, J. N., Boyle, J. A., Zatorre, R. J., & Jones-Gotman, M. (2008). Functional neuronal processing of body odors differs from that of similar common odors. *Cerebral Cortex, 18*(6), 1466-1474.

Lundström, J. N., Goncalves, M., Esteves, F., & Olsson, M. J. (2003). Psychological effects of subthreshold exposure to the putative human pheromone 4,16-androstadien-3-one. *Hormones And Behavior, 44*, 395-401.

Lundström, J. N., Gordon, A. R., Albrecht, J., Alden, E. C., & Boesveldt, S. (2010). Methods for building an inexpensive computer-controlled olfactometer for temporally precise behavioral experiments. *International Journal of Psychophysiology, 83*, 1-23.

Lundström, J. N., McClintock, M. K., & Olsson, M. J. (2006). Effects of reproductive state on olfactory sensitivity suggest odor specificity. *Biological Psychology, 71*(3), 244-247.

Lundström, J. N. & Olsson, M. J. (2005). Subthreshold amounts of social odorant affect mood, but not behavior, in heterosexual women when tested by a male, but not a female, experimenter. *Biological Psychology, 70*, 197-204.

Lundström, J. N., Olsson, M. J., & Larsson, M. (2000). *Effects of the putative pheromone 4,16-androstadien-3-one on psychological and psychophysiological variables: weak evidence.* Paper presented at the 22nd annual meeting of the association for chemoreception sciences, Sarasota.

Lundström, J. N., Olsson, M. J., Schaal, B., & Hummel, T. (2006). A putative social chemosignal elicits faster cortical responses than perceptually similar odorants. *Neuroimage, 30*, 1340-1346.

Maarse, H. (1991). *Volatile compounds in foods and beverages.* New York: Marcel Dekker.

Mackay, C. J. (1980). Measurement of mood and psychophysiological activity. In I. Martin & P. H. Venables (Eds.), *Techniques in Psychophysiology* (pp. 501-562). Chichester: Wiley.

Marsh, A. A., Ambady, N., & Kleck, R. E. (2005). The effects of fear and anger facial expressions on approach- and avoidance-related behaviors. *Emotion, 5*(1), 119-124.

McClintock, M. K. (1971). Menstrual Synchrony And Suppression. *Nature, 229*(5282), 244-&.

McClintock, M. K. (2003). Pheromones, Odors, and Vasanas: The Neuroendocrinology of Social Chemosignals in Humans and Animals. In D. W. Pfaff (Ed.), *Hormones, Brain, and Behavior* (Vol. 1, pp. 797-870): Academic Press.

Meijer, E. H., Smulders, F. T. Y., Merckelbach, H., & Wolf, A. G. (2007). The P300 is sensitive to concealed face recognition. *International Journal Of Psychophysiology, 66*(3), 231-237.

Melrose, D. R., Reed, H. C. B., & Patterson, R. I. (1971). Androgen Steroids Associated With Boar Odour As An Aid To Detection Of Oestrus In Pig Artificial Insemination. *British Veterinary Journal, 127*(10), 497-&.

Meredith, M. (2001). Human vomeronasal organ function: A critical review of best and worst cases. *Chemical Senses, 26*(4), 433-445.

Milinski, M. & Wedekind, C. (2001). Evidence for MHC-correlated perfume preferences in humans. *Behavioral Ecology, 12*(2), 140-149.

Monti-Bloch, L., Diaz-Sanchez, V., Jennings-White, C., & Berliner, D. L. (1998). Modulation of serum testosterone and autonomic function through stimulation of the male human vomeronasal organ (VNO) with Pregna-4,20-diene-3,6-dione. *Journal Of Steroid Biochemistry And Molecular Biology, 65*(1-6), 237-242.

Monti-Bloch, L. & Grosser, B. I. (1991). Effect Of Putative Pheromones On The Electrical-Activity Of The Human Vomeronasal Organ And Olfactory Epithelium. *Journal Of Steroid Biochemistry And Molecular Biology, 39*(4B), 573-582.

Monti-Bloch, L., Jennings-White, C., & Berliner, D. L. (1998). The human vomeronasal system - A review *Olfaction And Taste Xii - An International Symposium* (Vol. 855, pp. 373-389).

Monti-Bloch, L., Jennings-White, C., Dolberg, D. S., & Berliner, D. L. (1994). The Human Vomeronasal System. *Psychoneuroendocrinology, 19*(5-7), 673-686.

Morofushi, M., Shinohara, K., Funabashi, T., & Kimura, F. (2000). Positive relationship between menstrual synchrony and ability to smell 5 alpha-androst-16-en-3 alpha-ol. *Chemical Senses, 25*(4), 407-411.

Morris, J. S., Öhman, A., & Dolan, R. J. (1998). Conscious and unconscious emotional learning in the human amygdala. *Nature, 393*(6684), 467-470.

Moss, M., Hewitt, S., Moss, L., & Wesnes, K. (2008). Modulation of cognitive performance and mood by aromas of peppermint and ylang-ylang. *International Journal Of Neuroscience, 118*(1), 59-77.

Mühlberger, A., Wieser, M. J., Herrmann, M. J., Weyers, P., Tröger, C., & Pauli, P. (2009). Early cortical processing of natural and artificial emotional faces differs between lower and higher socially anxious persons. *Journal Of Neural Transmission, 116*(6), 735-746.

Mujica-Parodi, L. R.Strey, H. H.Frederick, B.Savoy, R.Cox, D.Botanov, Y.Tolkunov, D., et al. (2009). Chemosensory cues to conspecific emotional stress activate amygdala in humans. *Plos One, 4*(7).

Neumann, R., Hulsenbeck, K., & Seibt, B. (2004). Attitudes towards people with AIDS and avoidance behavior: Automatic and reflective bases of behavior. *Journal Of Experimental Social Psychology, 40*(4), 543-550.

Nixon, A., Mallet, A. I., & Gower, D. B. (1988). Simultaneous quantification of five odorous steroids (16-sndrostenes) in axillary hair of men. *The Journal of Steroid Biochemistry, 29*(5), 505-510.

Öhman, A. (2007). Has evolution primed humans to "beware the beast"? *Proceedings Of The National Academy Of Sciences Of The United States Of America, 104*(42), 16396-16397.

Öhman, A., Lundquist, D., & Esteves, F. (2001). The face in the crowd revisited: a threat advantage with schematic stimuli. *Journal of Personality and Social Psychology, 80*(3), 381-396.

Öhman, A. & Mineka, S. (2001). Fear, phobias and preparedness: toward an evolved module of fear and fear learning. *Psychological Review, 108*(3), 483-522.

Oinonen, K. A. & Mazmanian, D. (2002). To what extent do oral contraceptives influence mood and affect? *Journal Of Affective Disorders, 70*(3), 229-240.

Olofsson, J. K., Nordin, S., Sequeira, H., & Polich, J. (2008). Affective picture processing: An integrative review of ERP fndings. *Biological Psychology, 77*, 247-265.

Olsson, M. J., Lundström, J. N., Diamantopoulou, S., & Esteves, F. (2006). A putative female pheromone affects mood in men differently depending on social context. *European Review Of Applied Psychology, 56*(4), 279-284.

Oomura, Y., Aou, S., Koyama, Y., Fujita, I., & Yoshimatsu, H. (1988). Central Control Of Sexual-Behavior. *Brain Research Bulletin, 20*(6), 863-870.

Paller, K. A.Ranganath, C.Gonsalves, B.LaBar, K. S.Parrish, T. B.Gitelman, D. R.Mesulam, M. M., et al. (2003). Neural correlates of person recognition. *Learning & Memory, 10*(4), 253-260.

Pauli, P., Bourne, L. E., Diekmann, H., & Birbaumer, N. (1999). Cross-modality priming between odors and odor-congruent words. *American Journal Of Psychology, 112*(2), 175-186.

Pause, B. M., Adolph, D., Prehn-Kristensen, A., & Ferstl, R. (2009). Startle response potentiation to chemosensory anxiety signals in socially anxious individuals. *International Journal Of Psychophysiology, 74*(2), 88-92.

Pause, B. M., Krauel, K., Sojka, B., & Ferstl, R. (1998). Body odor evoked potentials: a new method to study the chemosensory perception of self and non-self in humans. *Genetica, 104*(3), 285-294.

Penn, D. J.Oberzaucher, E.Grammer, K.Fischer, G.Soini, H. A.Wiesler, D.Novotny, M. V., et al. (2007). Individual and gender fingerprints in human body odour. *Journal Of The Royal Society Interface, 4*(13), 331-340.

Polich, J. & Kok, A. (1995). Cognitive And Biological Determinants Of P300 - An Integrative Review. *Biological Psychology, 41*(2), 103-146.

Prehn-Kristensen, A.Wiesner, C.Bergmann, T. O.Wolff, S.Jansen, O.Mehdorn, H. M.Ferstl, R., et al. (2009). Induction of empathy by the smell of anxiety. *Plos One, 4*(6).

Prehn, A., Ohrt, A., Sojka, B., Ferstl, R., & Pause, B. M. (2004). Chemosensory anxiety signals augment the startle reflex. *Journal Of Psychophysiology, 18*(4), 210-210.

Preti, G., Cutler, W. B., Garcia, C. R., Huggins, G. R., & Lawley, H. J. (1986). Human Axillary Secretions Influence Womens Menstrual Cycles - The Role Of Donor Extract Of Females. *Hormones And Behavior, 20*(4), 474-482.

Preti, G. & Wysocki, C. J. (1999). Human pheromones: releasers or primers: fact or myth. In R. E. Johnston, D. Müller-Schwartze & P. Sorenseon (Eds.), *Advances in Chemical Communication in Vertebrates* (pp. 315-331). New York: Plenum Press.

Rikowski, A. & Grammer, K. (1999). Human body odour, symmetry and attractiveness. *Proceedings Of The Royal Society Of London Series B-Biological Sciences, 266*(1422), 869-874.

Rinck, M. & Becker, E. S. (2007). Approach and avoidance in fear of spiders. *Journal of Behavior Therapy and Experimental Psychiatry, 38*, 105-120.

Roberts, S. C., Gosling, L. M., Spector, T. D., Miller, P., Penn, D. J., & Petrie, M. (2005). Body odor similarity in noncohabiting twins. *Chemical Senses, 30*(8), 651-656.

Rotteveel, M. & Phaf, R. H. (2004). Automatic affective evaluation does not automatically predispose for arm flexion and extension. *Emotion, 4*(2), 156-172.

Rozenkrants, B., Olofsson, J. K., & Polich, J. (2008). Affective visual event-related potentials: Arousal, valence, and repetition effects for normal and distorted pictures. *International Journal Of Psychophysiology, 67*(2), 114-123.

Russell, M. J. (1976). Human Olfactory Communication. *Nature, 260*(5551), 520-522.

Russell, M. J., Switz, G. M., & Thompson, K. (1980). Olfactory Influences On The Human Menstrual-Cycle. *Pharmacology Biochemistry And Behavior, 13*(5), 737-738.

Sagiv, N. & Bentin, S. (2001). Structural encoding of human and schematic faces: Holistic and part-based processes. *Journal Of Cognitive Neuroscience, 13*(7), 937-951.

Salazar, L. T. H., Laska, M., & Luna, E. R. (2003). Olfactory Sensitivity for Aliphatic Esters in Spider Monkeys (Ateles geoffroyi). *Behavioral Neuroscience, 117*(6), 1142–1149.

Sato, W., Kochiyama, T., Yoshikawa, S., & Matsumura, M. (2001). Emotional expression boosts early visual processing of the face: ERP recording and its decomposition by independent component analysis. *Neuroreport, 12*(4), 709-714.

Savic, I., Berglund, H., Gulyas, B., & Roland, P. (2001). Smelling of odorous sex hormone-like compounds causes sex-differentiated hypothalamic activations in humans. *Neuron, 31*(4), 661-668.

Savic, I., Berglund, H., & Lindstrom, P. (2006). Brain response to putative pheromones in homosexual men. *Nordic Journal of Psychiatry, 60*(4), 327-327.

Savic, I., Berglund, H., & Lindström, P. (2005). Brain response to putative pheromones in homosexual men. *Proceedings Of The National Academy Of Sciences Of The United States Of America, 102*(20), 7356-7361.

Savic, I., Heden-Blomquist, E., & Berglund, H. (2009). Pheromone signal transduction in humans: what can be learned from olfactory loss. *Human Brain Mapping, 30*(9), 3057-3065.

Saxton, T. K., Lyndon, A., Little, A. C., & Roberts, S. C. (2008). Evidence that androstadienone, a putative human chemosignal, modulates women's attributions of men's attractiveness. *Hormones And Behavior, 54*(5), 597-601.

Schaal, B. & Porter, R. H. (1991). Microsmatic Humans Revisited - The Generation And Perception Of Chemical Signals. *Advances In The Study Of Behavior, 20*, 135-199.

Schank, J. C. (2006). Do human menstrual-cycle pheromones exist? *Human Nature-An Interdisciplinary Biosocial Perspective, 17*(4), 448-470.

Schupp, H. T., Junghöfer, M., Weike, A. I., & Hamm, A. O. (2003a). Attention and emotion: an ERP analysis of facilitated emotional stimulus processing. *Neuroreport, 14*(8), 1107-1110.

Schupp, H. T., Junghöfer, M., Weike, A. I., & Hamm, A. O. (2003b). Emotional facilitation of sensory processing in the visual cortex. *Psychological Science, 14*(1), 7-13.

Schupp, H. T., Öhman, A., Junghöfer, M., Weike, A. I., Stockburger, J., & Hamm, A. O. (2004). The facilitated processing of threatening faces: an ERP analysis. *Emotion, 4*(2), 189-200.

Shiferaw, B., Verrill, L., Booth, H., Zansky, S., Norton, D., Crim, S., & Henao, O. (2008, March 19). *Are there gender differences in food consumption? The FoodNet population survey, 2006-2007.* Paper presented at the International Conference on Emerging Infectious Diseases, Atlanta, GA.

Shinohara, K., Morofushi, M., Funabashi, T., & Kimura, F. (2001). Axillary pheromones modulate pulsatile LH secretion in humans. *Neuroreport, 12*(5), 893-895.

Shinohara, K., Morofushi, M., Funabashi, T., Mitsushima, D., & Kimura, F. (2000). Effects of 5 alpha-androst-16-en-3 alpha-ol on the pulsatile secretion of luteinizing hormone in human females. *Chemical Senses, 25*(4), 465-467.

Sobel, N. & Brown, W. M. (2001). The scented brain: pheromonal responses in humans. *Neuron, 31*(4), 512-514.

Solarz, A. (1960). Latency of instrumental responses as a function of compatibility with the meaning of eliciting verbal signs. *Journal of Experimental Psychology, 59*, 239-245.

Spehr, M., Gisselmann, G., Poplawski, A., Riffell, J. A., Wetzel, C. H., Zimmer, R. K., & Hatt, H. (2003). Identification of a testicular odorant receptor mediating human sperm chemotaxis. *Science, 299*(5615), 2054-2058.

Spielberger, C. S., Gorsuch, R. L., & Lushene, R. L. (1970). Manual for the State Trait Anxiety Inventory. Palo Alto, CA.: Consulting Psychological Press.

Stensaas, L. J., Lavker, R. M., Montibloch, L., Grosser, B. I., & Berliner, D. L. (1991). Ultrastructure Of The Human Vomeronasal Organ. *Journal Of Steroid Biochemistry And Molecular Biology, 39*(4B), 553-560.

Stern, K. & McClintock, M. K. (1998). Regulation of ovulation by human pheromones. *Nature, 392*(6672), 177-179.

Stoddart, D. M. (1990). *The scented ape: the biology and culture of human odor.* Cambridge: Cambridge University Press.

Strassmann, B. I. (1999). Menstrual synchrony pheromones: cause for doubt. *Human Reproduction, 14*(3), 579-580.

Sturm, W. & Willmes, K. (2001). On the functional neuroanatomy of intrinsic and phasic alertness. *NeuroImage, 14*, S76-S84.

Takami, S., Getchell, M. L., Chen, Y., Montibloch, L., Berliner, D. L., Stensaas, L. J., & Getchell, T. V. (1993). Vomeronasal Epithelial-Cells Of The Adult Human Express Neuron-Specific Molecules. *Neuroreport, 4*(4), 375-378.

Tales, A., Muir, J. L., Bayer, A., Jones, R., & Snowden, R. J. (2002). Phasic visual alertness in Alzheimer's disease and ageing. *Neuroreport, 13*(18), 2557-2560.

Thorne, F., Neave, N., Scholey, A., Moss, M., & Fink, B. (2002). Effects of putative male pheromones on female ratings of male attractiveness: Influence of oral contraceptives and the menstrual cycle. *Neuroendocrinology Letters, 23*(4), 291-297.

Thysen, B., Elliott, W. H., & Katzman, P. A. (1968). Identification Of Estra-1,3,5(10),16-Tetraen-3-Ol (Estratetraenol) From Urine Of Pregnant Women. *Steroids, 11*(1), 73-&.

Trotier, D., Eloit, C., Wassef, M., Talmain, G., Bensimon, J. L., Doving, K. B., & Ferrand, J. (2000). The vomeronasal cavity in adult humans. *Chemical Senses, 25*(4), 369-380.

Vandenbergh, J. G. (1969). Male Odor Accelerates Female Sexual Maturation In Mice. *Endocrinology, 84*(3), 658-&.

Villemure, C. & Bushnell, M. C. (2007). The effects of the steroid androstadienone and pleasant odorants on the mood and pain perception of men and women. *European Journal of Pain, 11*(2), 181-191.

Wallace, P. (1977). Individual Discrimination Of Humans By Odor. *Physiology & Behavior, 19*(4), 577-579.

Wang, J., Eslinger, P. J., Smith, M. B., & Yang, Q. X. (2005). Functional magnetic resonance imaging study of human olfaction and normal aging. *Journals Of Gerontology Series A-Biological Sciences And Medical Sciences, 60*(4), 510-514.

Wedekind, C., Seebeck, T., Bettens, F., & Paepke, A. J. (1995). Mhc-Dependent Mate Preferences In Humans. *Proceedings Of The Royal Society Of London Series B-Biological Sciences, 260*(1359), 245-249.

Weisfeld, G. E., Czilli, T., Phillips, K. A., Gall, J. A., & Lichtman, C. M. (2003). Possible olfaction-based mechanisms in human kin recognition and inbreeding avoidance. *Journal Of Experimental Child Psychology, 85*(3), 279-295.

Whalen, P. J., Rauch, S. L., Etcoff, N. L., McInerney, S. C., Lee, M. B., & Jenike, M. A. (1998). Masked presentations of emotional facial expressions modulate amygdala activity without explicit knowledge. *Journal Of Neuroscience, 18*(1), 411-418.

Whitten, W. (1999). Reproductive biology - pheromones and regulation of ovulation. *Nature, 401*(6750), 232-232.

Whitten, W. K., Bronson, F. H., & Greenstein, J. A. (1968). Estrus-inducing pheromone of male mice - transport by movement of air. *Science, 161*(3841), 584-&.

Wieser, M. J., Pauli, P., Reicherts, P., & Mühlberger, A. (2010). Don't look at me in anger! Enhanced processing of angry faces in anticipation of public speaking. *Psychophysiology, 47*(2), 271-280.

Wilke, K., Martin, A., Terstegen, L., & Biel, S. S. (2007). A short history of sweat gland biology. *International Journal of Cosmetic Science, 29*, 169-179.

Williams, M. A. & Mattingley, J. B. (2006). Do angry men get noticed? *Current Biology, 16*(11), R402-R404.

Wilson, D. A. & Stevenson, R. J. (2006). *Learning to smell: Olfactory perception from neurobiology to behavior.* Baltimore: Johns Hopkins University Press.

Wilson, E. O. & Bossert, W. H. (1963). Chemical communication among animals *Recent Progress in Hormone Research* (pp. 673-716).

Wilson, H. C. (1987). Female Axillary Secretions Influence Womens Menstrual Cycles - A Critique. *Hormones And Behavior, 21*(4), 536-546.

Wyart, C., Webster, W. W., Chen, J. H., Wilson, S. R., McClary, A., Khan, R. M., & Sobel, N. (2007). Smelling a single component of male sweat alters levels of cortisol in women. *Journal Of Neuroscience, 27*(6), 1261-1265.

Wyatt, T. D. (2003). *Pheromones and animal behaviour: Communication by smell and taste.* Cambridge: Cambridge University Press.

Wyatt, T. D. (2009). Fifty years of pheromones. *Nature, 457*(15), 262-263.

Wysocki, C. J. & Beauchamp, G. K. (1984). Ability To Smell Androstenone Is Genetically-Determined. *Proceedings Of The National Academy Of Sciences Of The United States Of America-Biological Sciences, 81*(15), 4899-4902.

Wysocki, C. J., Beauchamp, G. K., Schmidt, H. J., & Dorries, K. M. (1987). Changes In Olfactory Sensitivity To Androstenone With Age And Experience. *Chemical Senses, 12*(4), 710-710.

Yamazaki, K., Beauchamp, G. K., Curran, M., Bard, J., & Boyse, E. A. (2000). Parent-progeny recognition as a function of MHC odortype identity. *Proceedings Of The National Academy Of Sciences Of The United States Of America, 97*(19), 10500-10502.

Yamazaki, K., Curran, M., & Beauchamp, G. K. (1999). Fetal MHC odor types play a functional role in regulating social interactions. *Behavior Genetics, 29*(5), 375-375.

Zatorre, R. J., Jonesgotman, M., Evans, A. C., & Meyer, E. (1992). Functional Localization And Lateralization Of Human Olfactory Cortex. *Nature, 360*(6402), 339-340.

Zeng, X. N., Leyden, J. J., Brand, J. G., Spielman, A. I., McGinley, K. J., & Preti, G. (1992). An Investigation Of Human Apocrine Gland Secretion For Axillary Odor Precursors. *Journal Of Chemical Ecology, 18*(7), 1039-1055.

Zeng, X. N., Leyden, J. J., Lawley, H. J., Sawano, K., Nohara, I., & Preti, G. (1991). Analysis Of Characteristic Odors From Human Male Axillae. *Journal Of Chemical Ecology, 17*(7), 1469-1492.

Zhou, W. & Chen, D. (2008a). Chemosignal of Fear Modulates Fear Recognition in Ambiguous Facial Expressions. *Chemical Senses, 33*(8), S175-S175.

Zhou, W. & Chen, D. (2008b). Encoding Human Sexual Chemosensory Cues in the Orbitofrontal and Fusiform Cortices. *Journal Of Neuroscience, 28*(53), 14416-14421.

8 Appendix

8.1 Material of study I

8.1.1 Informed consent form

IRB Approval From: 5/27/2009 To: 5/26/2010

MONELL CHEMICAL SENSES CENTER
RESEARCH SUBJECT
INFORMED CONSENT FORM

Protocol Title:	Multimodal sensory processing: from behavior to neurons. Study 2
Principal Investigator:	Johan N. Lundström, Ph.D. 3500 Market Street, Philadelphia, PA 19104 267-519-4690
Emergency Contact:	24 hour access, call: (267) 519-4900

Why am I being asked to volunteer?

You are being invited to participate in a research study. Your participation is voluntary, which means you can choose whether or not you want to participate. If you choose not to participate, there will be no loss of benefits to which you are otherwise entitled. Before you can make your decision, you will need to know what the study is about, the possible risks and benefits of being in this study, and what you will have to do in this study. The research team is going to talk to you about the research study, and they will give you this consent form to read. You may also decide to discuss it with your family, friends, or family doctor. You may find some of the medical language difficult to understand. Please ask the study doctor and/or the research team about this form. If you decide to participate, you will be asked to sign this form.

What is the purpose of this research study?

In our everyday experience, our senses individually receive very different impressions of the world. From one object, our nose receives a smell, our ears receive a sound, and our eyes see an image. We understand quite well how our senses work by themselves; however, we do not know how they are able to interact to form a uniform percept of the world. The purpose of this study is to help us understand how the human brain processes these objects.

807759-AMD1CONT2009

Informed Consent Template © 2003-2006 Trustees of the University of Pennsylvania
Template Version: 14 Dec 2006

IRB Approval From: 5/27/2009 To: 5/26/2010

MULTIMODAL SENSORY PROCESSING

How long will I be in the study? How many other people will be in the study?

If you agree to volunteer, your participation in this study will require about 2 hours during one day. About 20 people will participate in this study.

What am I being asked to do?

First, you will complete this consent form. If you or a member of your family is an employee of the Monell Chemical Senses Center, you will be asked to complete an additional consent form. If you decide to sign all applicable consent forms and agree to volunteer for this study, we will ask you to complete a questionnaire to provide us with some basic demographic information and health data. After that, you will be presented either with smells that you can clearly perceive or with smells that will be harder to perceive. We will also test whether you are able to hear and see things by presenting sounds and having you read a few letters from a paper. We will then place a cap on your head containing small electrodes. These electrodes will measure your brain activity while you are presented with sounds, pictures, or smells. These may be presented by themselves or in different pairings with one other. After each presentation, you will be asked to answer a question regarding how you perceived them. Each task will be explained to you in detail before you have to perform any of them. Be sure to ask questions before each task if you are not sure what you should do or what you can expect.

What are the possible risks or discomforts?

You will smell chemical compounds. To the best of our knowledge, the compounds we present to you pose little risk under the conditions in which they will be presented. They are all smells that you may experience in everyday life. Nevertheless, any given person could have a negative reaction to any given compound – in other words, this study might involve currently unforeseen risks. If you feel unusual discomfort at any point, please inform the experimenter immediately. Certain conditions might increase your risk for negative effects. In particular, you should not volunteer if you have breathing problems or a history of sensitivity to chemicals. If you have any doubts, feel free to discuss them with the Principal Investigator of this study (Johan N. Lundstrom), who is listed on the first page of this form and/or with your physician (please bring this form to your doctor). In general, if you have any doubts, it is best not to volunteer for the study.

We know of no special risks of these study procedures for a pregnant woman or a fetus. However, if you are pregnant or suspect that you might be pregnant, you should not participate in this study. If you have any doubts, feel free to discuss them with the

Principal Investigator of this study (Johan N. Lundstrom), who is listed on the first page of this form and/or with your physician (please bring this form to your doctor). In general, if you have any doubts, it is best not to volunteer for the study.

If you have a medical emergency during the study you should go to the nearest emergency room. You may also contact the Principal Investigator or Emergency contact listed on page one of this form, or you may contact your own doctor. Be sure to tell the doctor or his/her staff that you are in a research study being conducted at the Monell Chemical Senses Center.

What if new information becomes available about the study?

During the course of this study, we may find more information that could be important to you. This includes information that, once learned, might cause you to change your mind about being in the study. We will notify you as soon as possible if such information becomes available.

What are the possible benefits of the study?

You are not expected to benefit personally from participating in this research study in any way. However, by agreeing to volunteer, you are helping to benefit society through advancing our understanding of the chemical senses.

What other choices do I have if I do not participate?

The only alternative to participation in this study is not to participate.

Will I be paid for being in this study?

Yes. You will receive financial compensation corresponding to the phase of the study you complete. You will receive 15 dollars, plus an additional 5 dollars to help with transportation costs, totaling 20 dollars upon completion.

Will I have to pay for anything?

You and/or your health insurance may be billed for the costs of medical care during this study if these expenses would have happened even if you were not in the study, or if your insurance agrees in advance to pay.

IRB Approval From: 5/27/2009 To: 5/26/2010

What happens if I am injured or hurt during the study?

If you have a medical emergency during the study you should go to the nearest emergency room. You may also contact the Principal Investigator or Emergency contact listed on page one of this form, or you may contact your own doctor. Be sure to tell the doctor or his/her staff that you are in a research study being conducted at the Monell Chemical Senses Center. Ask them to call the telephone numbers on the first page of this consent form for further instructions or information about your care.

Monell will not provide special services, free care, or compensation for any injuries resulting from this research. You, or your third party payer, if any, will be responsible for payment of any medical expenses that may result from injury incurred from participation in this study.

If you believe that you have suffered any injury as a result of participating in this research, you may contact the Principal Investigator of this study (Johan N. Lundstrom) at (267) 519-4690.

When is the Study over? Can I leave the Study before it ends?

This study is expected to end after all participants have been evaluated, and all information has been collected. This study may also be stopped at any time by the Monell Chemical Senses Center or the Food and Drug Administration (FDA).

In addition, the Principle Investigator may end your participation in the study if the Principle Investigator feels that 1) continued participation may be dangerous to your health or safety, or 2) you have not followed study directions. Such an action would not require your consent, but you will be informed if such a decision is made and of the reason for this decision.

You are free to leave the study anytime, for any reason. Withdrawal will not interfere with your future interactions in any way with this institution.

Who can see or use my information? How will my personal information be protected?

The investigator and staff involved with the study will keep any personal health information collected for the study strictly confidential. Every attempt will be made by the investigators to maintain all information collected in this study strictly confidential, except as may be required by court order or by law. Authorized representatives of the University of Pennsylvania Institutional Review Board (IRB), a committee charged with protecting the rights and welfare of research subjects, and the Monell Chemical Senses Center Human Subjects Committee (HSC) may be provided access to medical or

8.1.2 Onscreen instructions

Screen 1:

In the following task, you will be asked to react as fast as possible to pictures that appear on this screen. The pictures will either show persons with an ANGRY facial expression or persons with a HAPPY facial expression. You will use the joystick to respond. Do you have any questions? To CONTINUE, press the SPACE bar!

Screen 2:

Your task is to categorize the persons you see by their facial expressions. If you see a person with an ANGRY face, please PULL this ANGRY person closer to you by PULLING the joystick towards you. If you see a person with a HAPPY facial expression, please PUSH this HAPPY person away by PUSHING the joystick away from you. Do you have any questions? To CONTINUE, press the SPACE bar!

Screen 3:

Before starting, please be sure you are comfortable holding the joystick in the correct way You will use both hands, one hand to hold the base steady and the other hand to grasp the stick. Do you have any questions? To CONTINUE, press the SPACE bar!

Screen 4:

To start the task, you must wait for a crosshairs to appear on the screen. When you see it, press with the index finger the button on the joystick. Once the task begins, you must react as fast as possible without making mistakes! When you're ready, we will do a practice session. Do you have any questions? To CONTINUE, press the SPACE bar!

Practice trials

Screen 5:

Practice is over! Do you have any questions? When you're ready, press the SPACE bar to BEGIN the real task. If there are any problems during the task, please alert the experimenter.

Emotional response task

Screen 6:

Thank you! Now, the task will change slightly. You will still be reacting as quickly as possible to ANGRY and HAPPY faces using the joystick. The change involves how YOU should respond! To CONTINUE, press the SPACE bar.

Screen 7:

If you see a person with an ANGRY face, please PUSH this ANGRY person away by PUSHING the joystick away from you. If you see a person with a HAPPY facial expression, please PULL this HAPPY person closer by PULLING the joystick towards you. Remember to react as fast as possible without making mistakes. And, remember to use both hands – one hand to hold the base steady and the other hand to grasp the stick. When you're ready, we will do another practice session. Do you have any questions? To CONTINUE, press the SPACE bar.

Practice trials

Screen 8:

Practice is over! Do you have any questions? When you're ready, press the SPACE bar to BEGIN the real task. If there are any problems during the task, please alert the experimenter.

Emotional response task

Screen 9:

The task is now finished! Thank you very much. Please contact the experimenter now!

8.2 Material of study II

8.2.1 Informed consent form

UNIVERSITÄT WÜRZBURG

Lehrstuhl für Psychologie I - Prof. Dr. Paul Pauli
Biologische Psychologie, Klinische Psychologie und Psychotherapie

Prof. Dr. Paul Pauli
Marcusstr. 9-11
D-97070 Würzburg/Germany

Tel: +49 931 31 2842
Fax: +49 931 31 2733
e-Mail: pauli@psychologie.uni-wuerzburg.de

INFORMATION

Liebe Teilnehmerin,

Sie haben im Rahmen einer Doktorarbeit Gelegenheit an einer Untersuchung zur Verarbeitung emotionaler Bilder und Gesichter und deren Einfluss, sowie den Einfluss eines Duftstoffes auf körperliche Funktionen teilzunehmen.
Während Sie Gesichter und Bilder, die für Sie angenehm oder unangenehm sein können, betrachten, und zwei verschiedene Reaktionszeitaufgaben bearbeiten werden Hautleitfähigkeit, Herzrate, Atmung und die resultierende elektrische Aktivität des Gehirns aufgezeichnet.

Die Hautleitfähigkeit wird über zwei oberflächlich auf der Handinnenseite angebrachten Klebeelektroden aufgezeichnet. Die Herzrate wird mit zwei oberflächlich am Brustkorb angebrachten Klebeelektroden abgeleitet. Die Atmung wird mit einem Atemgürtel, der um den Brustkorb angebracht wird, erfasst. Zur Aufzeichnung Ihrer Hirnströme sind Sie an ein EEG-Gerät angeschlossen. Dies besteht aus einer Haube mit 28 Ableitungselektroden, die kleine, vom Gehirn produzierte Ströme registrieren. Um einen guten Kontakt der Elektroden zu erreichen werden die Kontaktstellen mit einem leitfähigen Gel bedeckt und leicht gerieben.
Die Verwendung des leitfähigen Gels kann eventuell zu Hautirritationen führen und macht eine anschließende Haarwäsche erforderlich.

Um den Einfluss eines Duftstoffes zu untersuchen wird Ihnen mit einem Wattepad eine kleine Flüssigkeitsmenge auf die Oberlippe aufgetragen. Es sind keine Nebenwirkungen des Duftstoffes, der auch in Parfümen verwendet wird, durch Einatmen über die Nasenschleimhaut bekannt.
Außerdem werden wir Sie bitten einige Speichelproben abzugeben und Fragebögen auszufüllen.

Die eigentliche Untersuchung findet am Computer statt. Nach Anbringen der Elektroden werden Sie zwei verschiedene Reaktionsaufgaben bearbeiten, die Ihnen zuvor am PC erklärt werden. Während Sie die Aufgaben bearbeiten sowie Bilder betrachten werden kontinuierlich die physiologischen Maße aufgezeichnet. Anschließend sollen Sie noch die Bilder nach verschiedenen Kriterien bewerten.

Insgesamt wird die Untersuchung ca. zwei Stunden in Anspruch nehmen. Es entstehen für die Teilnahme an dieser Studie keine Kosten für Sie. Die Teilnahme wird insgesamt mit 15 Euro vergütet.

Allgemeine Hinweise

Es ist wichtig, dass Sie auch auf allgemeine Prinzipien der Untersuchungen hingewiesen werden:

1. Die Teilnahme ist völlig freiwillig.
2. Sie selbst gewinnen keinen Nutzen aus der Untersuchung.
3. Sie können aus der Untersuchung jederzeit ausscheiden, ohne dass Sie deshalb einen Nachteil erleiden.
4. Alle erhobenen Daten werden anonymisiert und streng vertraulich nach den geltenden Datenschutzlinien behandelt.

Bei Fragen wenden Sie sich bitte an die Versuchsleiterin

Julius-Maximilians-
UNIVERSITÄT WÜRZBURG

Lehrstuhl für Psychologie I - Prof. Dr. Paul Pauli
Biologische Psychologie, Klinische Psychologie und Psychotherapie

Prof. Dr. Paul Pauli
Marcusstr. 9-11
D-97070 Würzburg/Germany

Tel: +49 931 31 2842
Fax: +49 931 31 2733
e-Mail: pauli@psychologie.uni-wuerzburg.de

Einverständniserklärung

Ich habe die oben aufgeführten Informationen gelesen und verstanden.

Ich bin darüber informiert worden, dass ich jederzeit aus der Untersuchung ausscheiden kann, ohne dass mir persönliche Nachteile entstehen.

Ich erkläre mich hiermit einverstanden, dass meine Daten anonymisiert zu Forschungszwecken verwendet werden.

Würzburg, den _____   Unterschrift: _____

Name (Druckschrift): _____
Anschrift (Druckschrift): _____
_____

Unterschrift der Versuchsleiterin: _____

8.2.2 Onscreen instructions

Vielen Dank, dass Sie am Experiment zur Erfassung der elektrischen Aktivität des Gehirns teilnehmen. Weiter mit der blauen Taste!

Bitte füllen Sie nun den ersten bereitliegenden Fragebogen aus. Wenn Sie fertig sind drücken Sie bitte die blaue Taste!

1ˢᵗ Mood and anxiety questionnaire

Im Folgenden sollen Sie eine Reaktionszeitaufgabe bearbeiten. In der Mitte des Bildschirms erscheint ein Kreuz. Wir bitten Sie sobald das Kreuz erscheint, so schnell wie möglich die blaue Taste zu drücken. Daraufhin verschwindet das Kreuz wieder. Sobald das nächste Kreuz erscheint drücken Sie bitte wieder so schnell wie möglich die blaue Taste. Insgesamt wird das Kreuz 10-mal erscheinen. Jetzt folgt ein Übungsdurchgang, damit Sie sich mit dem Ablauf vertraut machen können. Legen Sie hierfür bitte den rechten Zeigefinger auf die blaue Taste. Fertig? Dann drücken Sie bitte zum Start des Übungsdurchgangs die blaue Taste!

tonic alertness practice trials

Haben Sie den Ablauf verstanden? Wenn es Fragen gibt, wenden Sie sich bitte an die Versuchsleiterin. Die Aufgabe ist in 2 Blöcke aufgeteilt. Zwischen den Blöcken gibt es eine Pause von 15 Sekunden. Versuchen Sie in dieser Zeit zu entspannen. Bitte beachten Sie noch einmal: Versuchen Sie so schnell wie möglich auf das Kreuz zu reagieren. Wenn es keine Fragen mehr gibt, legen Sie jetzt ihren rechten Zeigefinger auf die blaue Taste und starten Sie die Aufgabe per blaue Tastendruck.

tonic alertness trials

Der 1. Block ist nun beendet. Bitte entspannen Sie sich und starten Sie jetzt die Pause mit der blauen Taste.

break

Die Pause ist nun beendet. Starten Sie jetzt bitte den 2. Block mit der blauen Taste.

tonic alertness trials

Die erste Aufgabe ist nun beendet. Als nächstes werden Sie eine weitere Reaktionsaufgabe bearbeiten. Diesmal erscheint kurz vor dem Kreuz, auf das Sie wieder so schnell wie möglich per Tastendruck reagieren sollen ein kleines Quadrat, das das Erscheinen des Kreuzes vorhersagt. Jetzt folgt ein Übungsdurchgang, damit Sie sich mit dem Ablauf vertraut machen können. Legen Sie hierfür bitte den rechten Zeigefinger auf die blaue Taste. Fertig? Dann drücken Sie bitte zum Start des Übungsdurchgangs die blaue Taste!

phasic alertness practice trials

Haben Sie den Ablauf verstanden? Wenn es Fragen gibt, wenden Sie sich bitte an die Versuchsleiterin. Die Aufgabe ist wieder in 2 Blöcke aufgeteilt. Zwischen den Blöcken gibt es eine Pause von 15 Sekunden. Versuchen Sie in dieser Zeit zu entspannen. Bitte beachten Sie noch einmal: Versuchen Sie so schnell wie möglich auf das Kreuz zu reagieren. Wenn es keine Fragen mehr gibt, legen Sie jetzt ihren rechten Zeigefinger auf die blaue Taste und starten Sie die Aufgabe per Tastendruck.

phasic alertness trials

Der 1. Block ist nun beendet. Bitte entspannen Sie sich und starten Sie jetzt die Pause mit der blauen Taste.

break

Die Pause ist nun beendet. Starten Sie jetzt bitte den 2. Block mit der blauen Taste.

phasic alertness trials

Bitte füllen Sie nun den zweiten bereitliegenden Fragebogen aus. Wenn Sie fertig sind drücken Sie bitte die blaue Taste!

2nd mood and anxiety questionnaires

Im Folgenden sollen Sie noch einmal die erste Reaktionszeitaufgabe bearbeiten. In der Mitte des Bildschirms erscheint wieder ein Kreuz. Wir bitten Sie wieder sobald das Kreuz erscheint, so schnell wie möglich die blaue Taste zu drücken. Starten Sie die Aufgabe bitte mit der blauen Taste!

tonic alertness trials

Der 1. Block ist nun beendet. Bitte entspannen Sie sich und starten Sie jetzt die Pause mit der blauen Taste.

break

Die Pause ist nun beendet. Starten Sie jetzt bitte den 2. Block mit der blauen Taste.

tonic alertness trials

Als nächstes werden Sie wieder die zweite Reaktionsaufgabe bearbeiten. Diesmal erscheint kurz vor dem Kreuz wieder ein Quadrat als Hinweisreiz. Reagieren Sie bitte so schnell wie möglich auf das Kreuz per Tastendruck. Starten Sie die Aufgabe bitte mit der blauen Taste!

Trials phasic alertness

Der 1. Block ist nun beendet. Bitte entspannen Sie sich und starten Sie jetzt die Pause mit der blauen Taste.

break

Die Pause ist nun beendet. Starten Sie jetzt bitte den 2. Block mit der blauen Taste.

phasic alertness trials

Bitte füllen Sie nun den dritten bereitliegenden Fragebogen aus. Wenn Sie fertig sind drücken Sie bitte die blaue Taste!

3rd mood and anxiety questionnaires

Bitte geben Sie jetzt die zweite Speichelprobe in das bereitstehende Gefäß ab. Wenn Sie fertig sind drücken Sie bitte die blaue Taste!

2nd saliva collection

Bitte füllen Sie nun den vierten bereitliegenden Fragebogen aus. Wenn Sie fertig sind drücken Sie bitte die blaue Taste!

4th mood and anxiety questionnaire

Bitte nehmen Sie jetzt die bereitliegende Einwegspritze, geben die Flüssigkeit auf den Wattepad und applizieren Sie nun die Flüssigkeit auf ihre Oberlippe. Wenn Sie fertig sind drücken Sie bitte die blaue Taste!

2nd compound application

Ihre Aufgabe ist es nun wieder, die folgenden Bilder anzuschauen. Diesmal erscheint vor jedem Bild ein Kreuz in Mitte des Bildschirms. Bitte blicken Sie auf dieses bis das eigentliche Bild erscheint. Weiter mit der blauen Taste!

Bitte füllen Sie nun den fünften bereitliegenden Fragebogen aus. Wenn Sie fertig sind drücken Sie bitte die blaue Taste!

5th mood and anxiety questionnaire

Wenn Sie noch Fragen haben, sagen Sie bitte Bescheid, die Versuchsleiterin wird dann zu Ihnen kommen. Ansonsten starten Sie den Versuch mit der blauen Taste!

presentation of faces and scenes

Bitte füllen Sie nun den sechsten bereitliegenden Fragebogen aus. Anschließend bitte die blaue Taste drücken!

6th mood and anxiety questionnaires

Nun werden Ihnen verschiedene Personen angezeigt. Sie sollen diese nach vier verschiedenen Kriterien bewerten. Benutzen Sie hierfür bitte die obere Zahlenreihe Ihrer Tastatur. Es gibt dabei keine richtigen oder falschen Antworten. Ihre ehrliche und genaue gefühlsmäßige Einschätzung ist von Interesse. Versuchen Sie bitte so schnell und spontan wie möglich zu antworten. Eine Zeitbegrenzung besteht aber nicht. Weiter mit der blauen Taste!

Jede Person erscheint vier Mal mit jeweils einer von vier verschiedenen Bewertungsskalen. Lesen Sie sich bitte jede Skala vor der Bewertung genau durch! Starten Sie die Bewertung bitte jetzt mit der blauen Taste!

face rating

Nun werden Ihnen verschiedene Bilder angezeigt. Sie sollen diese auf zwei verschiedenen Skalen von 1 bis 9 bewerten. Benutzen Sie hierfür bitte wieder die obere Zahlenreihe Ihrer Tastatur. Es gibt dabei keine richtigen oder falschen Antworten. Ihre ehrliche und genaue gefühlsmäßige Einschätzung ist von Interesse. Versuchen Sie bitte so schnell und spontan wie möglich zu antworten. Eine Zeitbegrenzung besteht aber nichtWeiter mit der blauen Taste!

Jedes Bild erscheint zwei Mal, mit einer der beiden Bewertungsskalen. Lesen Sie sich bitte jede Skala vor der Bewertung genau durch! Starten Sie jetzt die Bewertung bitte mit der blauen Taste!");

scene rating

Bitte füllen Sie nun den siebten bereitliegenden Fragebogen aus. Anschließend bitte die blaue Taste drücken!

7th mood and anxiety questionnaires

Bitte geben Sie jetzt die dritte Speichelprobe in das bereitstehende Gefäß ab. Wenn Sie fertig sind drücken Sie bitte die blaue Taste!

3rd saliva collection

Bitte füllen Sie nun den achten bereitliegenden Fragebogen und die Nachbefragung aus.

8th mood and anxiety questionnaires

Vielen Dank für Ihre Teilnahme! Die Aufgabe ist nun beendet. Bitte kontaktieren sie die Versuchsleiterin!

finish

8.2.3 Plots of mood reactions

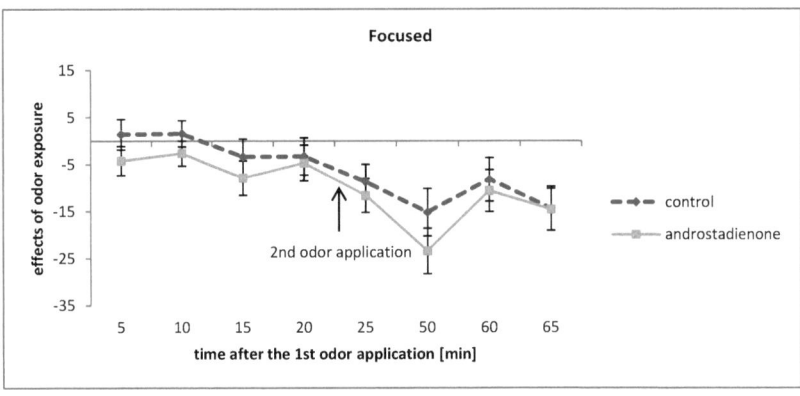

Figure 26. Mean odor effects on feeling focused ± SEM in control (dark) and androstadienone (light) groups.

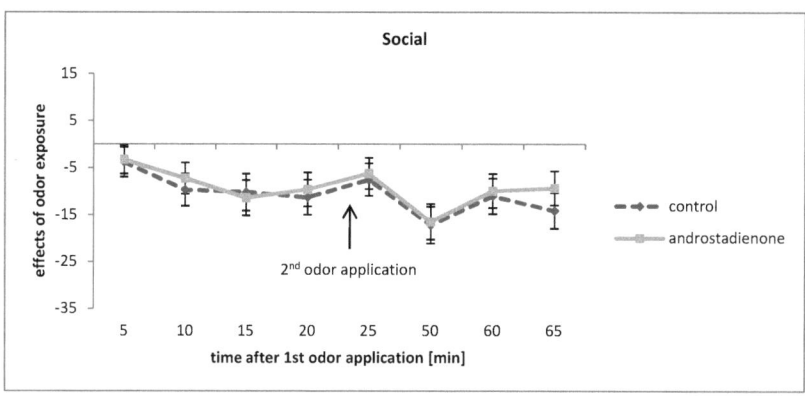

Figure 27. Mean odor effects on feeling social ± SEM in control (dark) and androstadienone (light) groups.

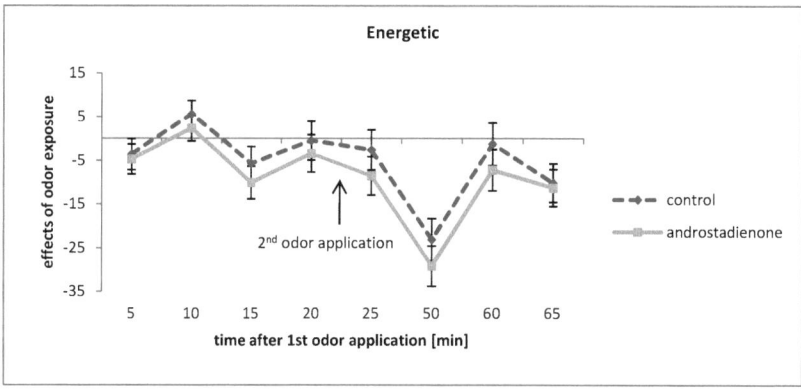

Figure 28. Mean odor effects on feeling energetic ± SEM in control (dark) and androstadienone (light) groups.

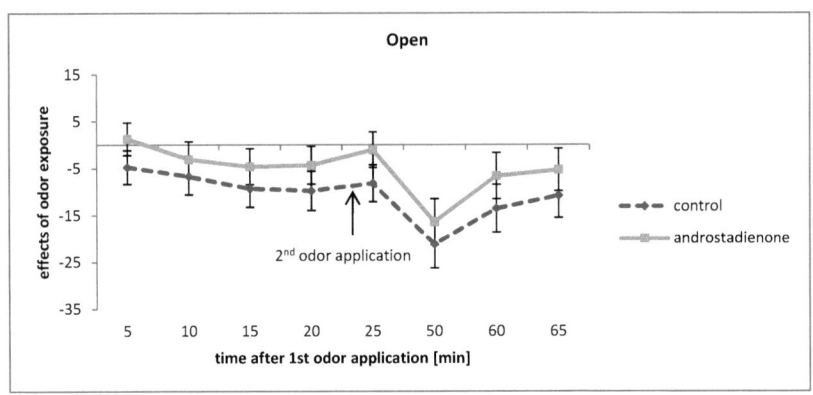

Figure 29. Mean odor effects on feeling open ± SEM in control (dark) and androstadienone (light) groups.

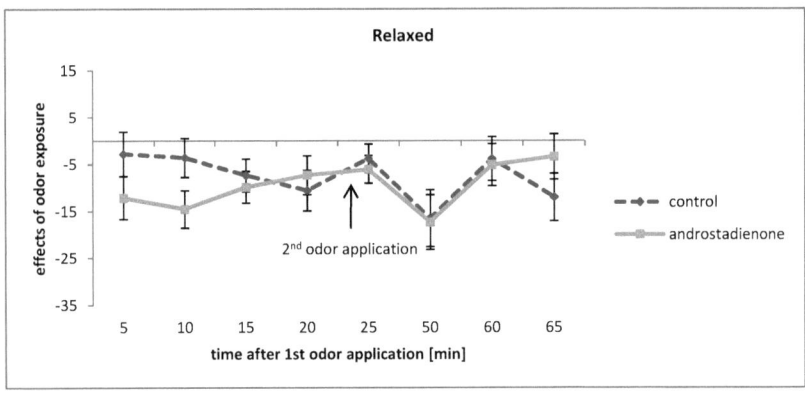

Figure 30. Mean odor effects on feeling relaxed ± SEM in control (dark) and androstadienone (light) groups.

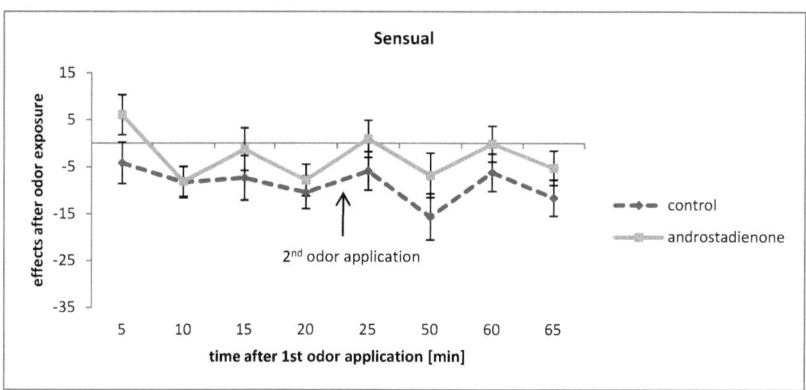

Figure 31. Mean odor effects on the feeling of being sensual ± SEM in control (dark) and androstadienone (light) groups.

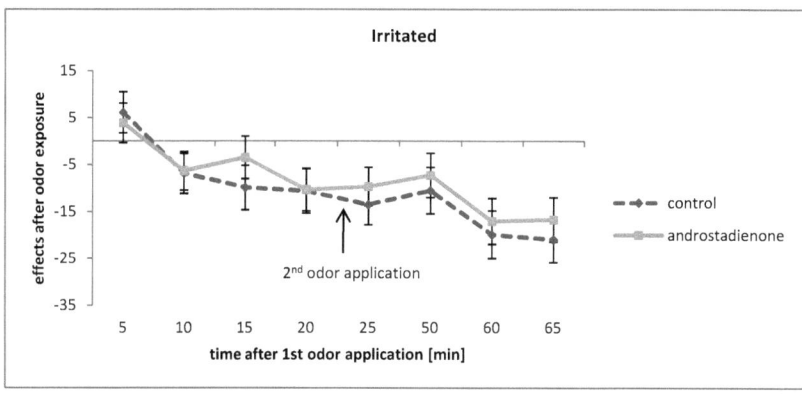

Figure 32. Mean odor effects on feeling irritated ± SEM in control (dark) and androstadienone (light) groups.

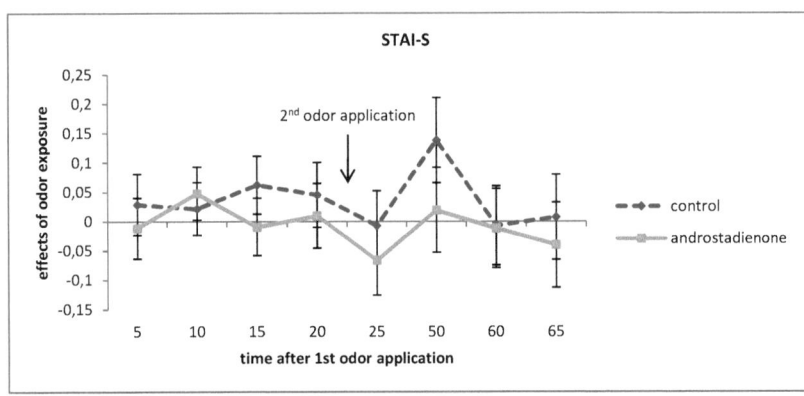

Figure 33. Mean odor effects on state anxiety ± SEM in control (dark) and androstadienone (light) groups.

8.3 Material of study III

8.3.1 Informed consent form

Approved from 12/10/2009 Approved to 12/09/2010

MONELL CHEMICAL SENSES CENTER
RESEARCH SUBJECT
INFORMED CONSENT FORM

Protocol Title:	Cortical measures of sensory function and integration
Principal Investigator:	Johan N. Lundstrom, Ph.D. 3500 Market Street, Philadelphia, PA 19104 267-519-4690
Emergency Contact:	24 hour access, call: (267) 519-4900

Why am I being asked to volunteer?

You are being invited to participate in a research study. Your participation is voluntary, which means you can choose whether or not you want to participate. If you choose not to participate, there will be no loss of benefits to which you are otherwise entitled. Before you can make your decision, you will need to know what this study is about, the possible risks and benefits of being in this study, and what you will have to do in this study. The research team is going to talk to you about the research study, and they will give you this consent form to read. You may also decide to discuss it with your family, friends, or family doctor. You may find some of the medical language difficult to understand. Please ask the study doctor and/or the research team about this form. If you decide to participate, you will be asked to sign this form.

What is the purpose of this research study?

We are investigating how the human brain processes different combinations of stimuli. By exploring differences in brain activity in response to combinations of sights, sounds, and chemosensory stimuli, we hope to reach a better understanding of how the brain works. For example, we may compare your brain's response to a picture to its response to a picture and a smell, together. In particular, we are interested in how the brain helps us detect, identify, and differentiate types of sensory input.

How long will I be in the study? How many other people will be in the study?

You and the experimenter have already discussed the types of stimuli you will experience and the tasks you will complete today. The combinations of stimuli you receive and tasks you perform will

ORIG

Informed Consent Template © 2005-2006 Trustees of the University of Pennsylvania
Template Version: 14 Dec 2006

Approved from 12/10/2009 Approved to 12/09/2010
CORTICAL MEASURES OF SENSORY FUNCTION AND INTEGRATION

determine the length of your participation today and whether you will need to come back another day. Approximately 150 people will participate in this study.

In the table below, the experimenter has indicated the number of sessions and the date(s) on which you are being asked to participate.

Please initial and date the line(s) corresponding to the session(s) you will participate in today. (If you are unsure, please ask.)

Session(s)	Date(s)	SUBJECT	EXPERIMENTER
Session 1			
Session 2 (if necessary)			
Session 3 (if necessary)			
Session 4 (if necessary)			

What am I being asked to do?

You will start by filling out some paperwork. First, you will complete this consent form. If you or a member of your family is an employee of the Monell Chemical Senses Center, you will be asked to complete an additional consent form. If you decide to sign all applicable consent forms and agree to volunteer for the study, we will ask you to complete a questionnaire to provide some basic demographic information and health data about yourself.

To begin, we will place a cap on your head containing small electrodes. These electrodes will measure your brain activity while you experience various types of stimuli. During the study you may experience any of the following stimuli: sounds, sights, smells, or tastes. These may be presented by themselves or in different pairings with one another.

You will also be asked to complete various tasks while we record your brain activity depending on which stimuli you experience. Each task will be explained to you in detail before you have to perform it. Be sure to ask questions before each task if you are not sure what you should do or what you can expect.

Below, please read the brief description(s) of the task(s) you will perform. (You will be given more detailed instructions before performing any of these tasks.)

(1) Emotional ratings – You may be asked to give ratings of your current emotional state and/or ratings of your general emotional state.

(2) Perceptual ratings – You may be presented with stimuli and asked to rate multiple characteristics, such as the strength or pleasantness, of each one.

(3) Sensory discrimination – You may be presented with sets of stimuli and asked to tell us which one in each set is different from the others.

(4) Sensory sensitivity – You may be presented with sets of stimuli, some weak and some strong, and asked to tell us which one in each set has a smell, taste, image, and/or sound.

(5) Sensory identification – You may be presented with stimuli and asked to tell what they smell, taste, look, and/or sound like.

(6) Reaction speed – You may be presented with stimuli and asked to react to the smells, tastes, images, and/or sounds as fast as possible by pressing a button.

(7) Psychophysiological measures – You may be presented with stimuli while we record your body's responses to smells, tastes, images, and/or sounds. We may record your skin conductance with electrodes on your fingertips, skin temperature with a probe taped to your hand, and/or heart rate with a clip attached to your finger. (All of these devices measure natural changes in your body's rhythms, and all are built for safe use with human research subjects.)

Although we cannot tell you how well you perform during testing, if you are interested, we can tell you how well you performed after you have completed all the required tasks.

Remember, you will be given detailed instructions before each task, but please be sure to ask questions if you are not sure what to do or what to expect.

Finally, please remember that you are free to leave the study anytime, for any reason.

What are the possible risks or discomforts?

In addition to the sights and sounds, you may smell and taste chemical compounds, so there are a few minor risks associated with those tasks. To the best of our knowledge, the compounds we present to you pose little risk under the conditions of this study. They are all part of household products that you may experience in everyday life. Nevertheless, any given person could have a negative reaction to any given chemical compound – in other words, this study might involve currently unforeseen risks. **If you feel unusual discomfort at any point, please inform the experimenter immediately.**

One task you may perform requires rating multiple dimensions of your current and/or general emotional state, and there is a possibility that this task may make you uncomfortable. If you do become uncomfortable and do not wish to complete this task, please remember that you are free to leave the study anytime, for any reason.

In general, certain conditions might increase your risk for negative effects. You should not participate in this study if you have known heart problems or a history of sensitivity to chemicals. We know of no special risks posed by this study for pregnant women or for the fetus. However, if you are pregnant or suspect that you might be pregnant, you should not participate in this study.

Approved from 12/10/2009 Approved to 12/09/2010
CORTICAL MEASURES OF SENSORY FUNCTION AND INTEGRATION

If you have any doubts, feel free to discuss them with the Principal Investigator of this study (Dr. Johan N. Lundstrom), who is listed on the first page of this form and/or with your physician. (In that case, please bring this form to your doctor.) In general, if you have any doubts, it is best not to participate in this study.

If you have a medical emergency during the study, you should go to the nearest emergency room. You may also contact the Principal Investigator or Emergency contact listed on the first page of this form and/or with your physician. In that case, please bring this form to your doctor and be sure to tell his/her staff that you are in a research study being conducted at the Monell Chemical Senses Center.

What if new information becomes available about the study?

During the course of this study, we may find more information that could be important to you. This includes information that, once learned, might cause you to change your mind about being in the study. We will notify you as soon as possible if such information becomes available.

What are the possible benefits of the study?

You are not expected to benefit personally from participating in this research study in any way. However, by agreeing to volunteer, you are helping to benefit society through advancing our understanding of the chemical senses.

What other choices do I have if I do not participate?

The only alternative to participation in this study is not to participate.

Will I be paid for being in this study?

Yes. You will be compensated financially upon completion of the EEG recording session(s). Because there are various conditions in this study, and each condition varies in length, you will be compensated according to the length of your participation.

You will be paid $15 USD for the first half hour; this includes $5 USD to compensate for transportation costs and $10 for fitting of the cap and electrodes. You will be paid $10 USD for each additional half hour of testing. Any payment above the $5 USD to compensate for transportation costs is contingent upon completion of a full testing session.

The maximum amount you will receive for participation is $_____ for _____ hours.

Will I have to pay for anything?

You and/or your health insurance may be billed for the costs of medical care incurred during this study if either those costs would have been incurred even if you were not participating in the study, or if your insurance agrees to pay in advance.

Approved from 12/10/2009 Approved to 12/09/2010
CORTICAL MEASURES OF SENSORY FUNCTION AND INTEGRATION

What happens if I am injured or hurt during the study?

If you have a medical emergency during the study, you should go to the nearest emergency room. You may also contact the Principal Investigator or Emergency contact listed on the first page of this form and/or with your physician. In that case, please bring this form to your doctor and be sure to tell his/her staff that you are in a research study being conducted at the Monell Chemical Senses Center. Ask them to call the telephone numbers on the first page of this consent form for further instructions or information about your care.

Monell will not provide special services, free care, or compensation for any injuries resulting from this research. You, or your third party payer, if any, will be responsible for payment of any medical expenses that may result from injury incurred from participation in this study.

If you believe that you have suffered any injury as a result of participating in this study, you may contact the Principal Investigator (Dr. Johan N. Lundstrom) at (267) 519-4690.

When is the Study over? Can I leave the Study before it ends?

This study is expected to end after all participants have been evaluated, and all information has been collected. This study may also be stopped at any time by the Monell Chemical Senses Center or the Food and Drug Administration (FDA).

In addition, the Principal Investigator may end your participation in the study if the Principal Investigator feels that 1) continued participation may be dangerous to your health or safety, or 2) you have not followed the experimenter's instructions. Such an action would not require your consent, but you will be informed if such a decision is made and of the reason for this decision.

You are free to leave the study anytime, for any reason. Withdrawal will not interfere in any way with your future interactions with this institution.

Who can see or use my information? How will my personal information be protected?

The investigator and staff involved with the study will keep any personal health information collected for the study strictly confidential. Every attempt will be made by the investigators to maintain strict confidentiality for all information collected in this study, except under conditions as may be required by court order or by law. Authorized representatives of the University of Pennsylvania Institutional Review Board (IRB), a committee charged with protecting the rights and welfare of research subjects, and the Monell Chemical Senses Center Human Subjects Committee (HSC) may be provided access to medical or research records that identify you by name. You will never be identified by name in any publication or in any presentations resulting from this research study.

> *Approved from 12/10/2009 Approved to 12/09/2010*
> CORTICAL MEASURES OF SENSORY FUNCTION AND INTEGRATION
>
> **Who can I call with questions, complaints, or if I'm concerned about my rights as a research subject?**
>
> If you have questions, concerns, or complaints regarding your participation in this research study, or if you have any questions about your rights as a research subject, you should speak with the Principal Investigator listed on page one of this form (Dr. Johan N. Lundstrom, 267-519-4690). If a member of the research team cannot be reached or if you want to talk to someone other than those working on the study, you may contact the Office of Regulatory Affairs at the University of Pennsylvania (215-898-2614) with any questions, concerns, or complaints.
>
> > When you sign this form, you are volunteering to take part in this research study. This means that you have read the consent form, your questions have been answered, and you have decided to volunteer. Your signature also means that you are permitting the Monell Chemical Senses Center to use your personal health information collected about you for research purposes within our institution. You are also allowing the Monell Chemical Senses Center to disclose that personal health information to outside organizations or people involved with the operations of this study.
>
> A copy of this consent form will be given to you.
>
> _____ _____ _____
> SUBJECT NAME (PRINT)   SUBJECT SIGNATURE   DATE
>
> _____ _____ _____
> EXPERIMENTER NAME (PRINT) EXPERIMENTER SIGNATURE DATE
>
> ORIG   6 of 6
> *Informed Consent Template © 2005-2006 Trustees of the University of Pennsylvania*
> *Template Version: 14 Dec 2006*

8.3.2 Onscreen instructions

Screen 1:

Welcome to the Monell Chemical Senses Center! Thank you for participating in this study. Please press the space bar to continue!

Screen 2:

During this task you will view cartoon faces on the screen! After each face you will be asked to identify the emotion of the presented face as quickly and accurately as possible. Press the space bar to continue!

Screen 3:

To respond, you will use the numbers on the keyboard in your lap. If you see an angry face please press button number 1. If you see a happy face please press button number 2. If you see a neutral face please press button number 3. Press the space bar to continue!

Screen 4:

After each face you will see a question on the screen. You have 2 seconds to answer the question. If you don't press a button within 2 seconds or you press an incorrect button you will repeat the trial. Press the space bar to continue!

Screen 5:

Please do NOT press a button during the picture is on the screen! You should always respond once the question appears on the screen. Press the space bar to continue!

Screen 6:

For each answer you should use the index finger of your dominant hand, without looking at the keyboard. To start place your index finger on button number 2. Then, answer the question by pressing the correct button and moving your finger left or right accordingly. After your answer return to the start position on button number 2. Remember; do not look at the keyboard during your response.

Any questions? Press the space bar to continue!

Screen 7:

Now, there will be a short practice session. Please keep in mind: 1 = angry, 2 = happy, 3 = neutral; Try to respond as quickly and accurately as possible. Remember to use the index finger of your dominant hand without looking at the keyboard. If you have any questions, please contact the experimenter. If not, please put your index finger on button number 2. To start the practice session, press the space bar!

Practice trials

Screen 8:

You have now completed the practice session. Please contact the experimenter.

To start the task, press the space bar now.

Real task

Final Screen:

Thank you! The task is finished! Please contact the experimenter.

8.4 Rating scales of study I and study III

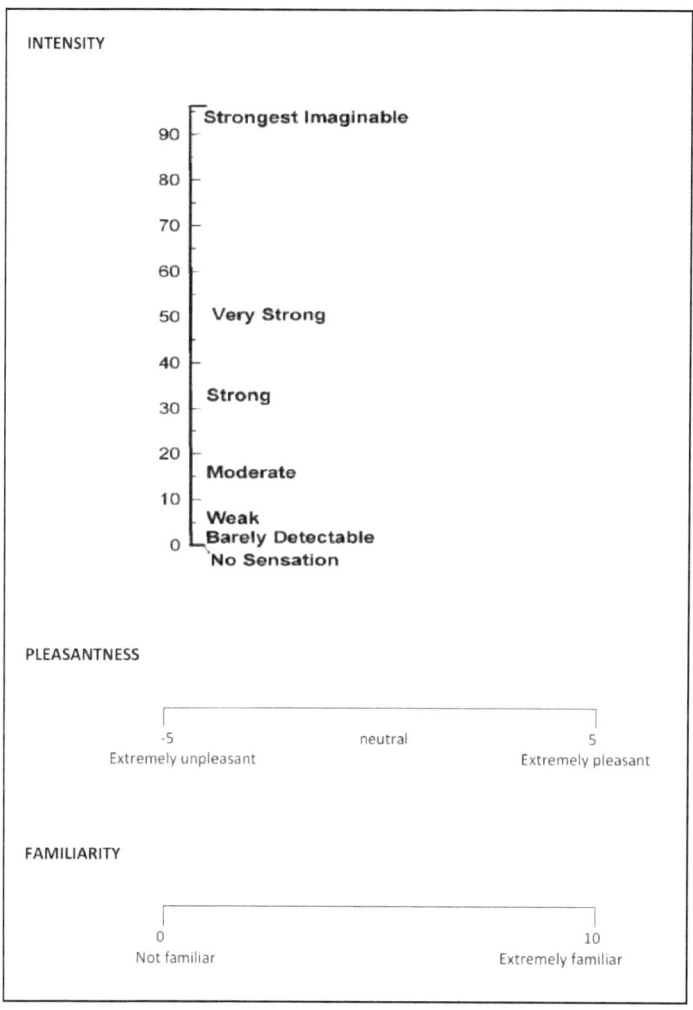

8.5 Publications

8.5.1 Articles

Frey, M.C.M., Weyers, P., Pauli, P., Mühlberger, A., (2012). Androstadienone in motor reactions of men and women towards angry faces. *Perceptual and Motor Skills,* 114(3), 807-825.

Mühlberger, A., Wieser, M.J., Gerdes, A.B.M., **Frey, M.C.M.**, Weyers, P., Pauli, P., (2011). Stop looking angry and smile, please: start and stop of the very same facial expression differentially activate threat- and reward-related brain networks. *SCAN,* 6, 321-329.

8.5.2 Conference proceedings

Frey, M.C.M., Lundström, J.N., Weyers, P., Pauli, P., Mühlberger, A., 2010. Androstadienone affects attention-related reactions. *32nd annual meeting of the Association for Chemoreception Sciences (AChemS)*, St. Pete Beach, FL, US.

Frey, M.C.M., Weyers, P., Pauli, P., Mühlberger, A., 2010. Effects and mechanisms of a putative human pheromone. *Spring School "Psychopathology of Emotions"*, Schweinfurt, Germany.

Wieser, M.J., Mühlberger, A., Gerdes, A.B.M., **Frey, M.C.M**, Weyers, P., Pauli, P., 2009. Happy end: the offset of angry facial expressions activates the human reward system. *Psychophysiology*, (46), S64.

Frey, M.C.M., Weyers, P., Pauli, P., Mühlberger, A., 2009. Human pheromones – fact or fantasy? *Neurobiology PhD student workshop*, Wuerzburg, Germany.

Frey, M.C.M., Weyers, P., Pauli, P., Mühlberger, A., 2009. A human pheromone as safety signal? *Keystone Symposium on Chemical Senses: Receptors and Circuits*, Tahoe City, CA, US

Frey, M.C.M., Weyers, P., Pauli, P., Mühlberger, A., 2008. Effects and mechanisms of putative human pheromones. *Summer School "Biopsychology of Emotions"*, Plankstetten, Germany.

Frey, M.C.M., Weyers, P., Pauli, P., Mühlberger, A., 2008. Facial reactions to social and non-social stimuli. *12th European Conference on Facial Expression*, Geneva, Switzerland.

Frey, M.C.M., Weyers, P., Pauli, P., Mühlberger, A., 2008. Effects and mechanisms of putative human pheromones. *5th Brain and Behavior Days "Comparative Research on Emotion Processing"*, Bronnbach, Germany.

Weyers, P., **Frey, M.C.M.**, Likowski, K., Wieser, M.J., Pauli, P., Mühlberger, A., 2007. A direct comparison of facial reactions to social and non-social emotional stimuli. *Psychophysiology*, (44), S36.

Mühlberger, A., **Frey, M.C.M.**, Likowski, K., Wieser, M.J., Pauli, P., Weyers, P., 2007. A direct comparison of central nervous processing of social and non-social emotional stimuli. *Psychophysiology* (44), S36.

i want morebooks!

Buy your books fast and straightforward online - at one of world's fastest growing online book stores! Environmentally sound due to Print-on-Demand technologies.

Buy your books online at
www.get-morebooks.com

Kaufen Sie Ihre Bücher schnell und unkompliziert online – auf einer der am schnellsten wachsenden Buchhandelsplattformen weltweit! Dank Print-On-Demand umwelt- und ressourcenschonend produziert.

Bücher schneller online kaufen
www.morebooks.de

 VDM Verlagsservicegesellschaft mbH
Heinrich-Böcking-Str. 6-8   Telefon: +49 681 3720 174   info@vdm-vsg.de
D - 66121 Saarbrücken   Telefax: +49 681 3720 1749 www.vdm-vsg.de

Printed by Books on Demand GmbH, Norderstedt / Germany